T0174851

Maintainability, Maintenance, and Reliability for Engineers

Maintainability, Maintenance, and Reliability for Engineers

B.S. Dhillon

CRC Press
Taylor & Francis Group
Boca Raton London New York

CRC Press is an imprint of the
Taylor & Francis Group, an **informa** business

CRC Press
Taylor & Francis Group
6000 Broken Sound Parkway NW, Suite 300
Boca Raton, FL 33487-2742

First issued in paperback 2019

© 2006 by Taylor & Francis Group, LLC
CRC Press is an imprint of Taylor & Francis Group, an Informa business

No claim to original U.S. Government works

ISBN-13: 978-0-8493-7243-8 (hbk)
ISBN-13: 978-0-367-39100-3 (pbk)

Library of Congress Card Number 2005056882

This book contains information obtained from authentic and highly regarded sources. Reasonable efforts have been made to publish reliable data and information, but the author and publisher cannot assume responsibility for the validity of all materials or the consequences of their use. The authors and publishers have attempted to trace the copyright holders of all material reproduced in this publication and apologize to copyright holders if permission to publish in this form has not been obtained. If any copyright material has not been acknowledged please write and let us know so we may rectify in any future reprint.

Except as permitted under U.S. Copyright Law, no part of this book may be reprinted, reproduced, transmitted, or utilized in any form by any electronic, mechanical, or other means, now known or hereafter invented, including photocopying, microfilming, and recording, or in any information storage or retrieval system, without written permission from the publishers.

For permission to photocopy or use material electronically from this work, please access www.copyright. com (http://www.copyright.com/) or contact the Copyright Clearance Center, Inc. (CCC), 222 Rosewood Drive, Danvers, MA 01923, 978-750-8400. CCC is a not-for-profit organization that provides licenses and registration for a variety of users. For organizations that have been granted a photocopy license by the CCC, a separate system of payment has been arranged.

Trademark Notice: Product or corporate names may be trademarks or registered trademarks, and are used only for identification and explanation without intent to infringe.

Library of Congress Cataloging-in-Publication Data

Dhillon, B. S. (Balbir S.), 1947-
    Maintainability, maintenance, and reliability for engineers / B.S. Dhillon.
        p. cm.
    Includes bibliographical references and index.
    ISBN 0-8493-7243-7 (alk. paper)
    1. Systems engineering--Management. I. Title.

TA168.D53 2006
620.001'171--dc22   2005056882

Visit the Taylor & Francis Web site at
http://www.taylorandfrancis.com

and the CRC Press Web site at
http://www.crcpress.com

This book is affectionately dedicated to Professor
Alexander S. Krausz
for hiring me at the University of Ottawa
and inspiring me to write journal articles and books.

Preface

In today's competitive global economy, it is becoming clear that only those nations that lead in technology will lead the world. Within the framework of technology, we include research, development, and manufacturing, as well as maintainability, maintenance, and reliability. Various studies indicate that for many large and sophisticated products and systems, maintenance and support costs account for as much as 60 to 75% of their life cycle costs.

Needless to say, maintainability, maintenance, and reliability of such systems and products have become more important than ever. Global competition and other associated factors are forcing manufacturers to produce highly reliable and easily maintainable engineering products.

This means there is a definite need for the maintainability, maintenance, and reliability professionals and associated personnel to work together during the product and system design, and other phases. To achieve this goal effectively it is absolutely essential that they have, to a certain degree, an understanding of each other's disciplines.

To the best of the author's knowledge, no book covers the topics of maintainability, maintenance, and reliability within its framework. Thus, to gain knowledge of each other's specialties, these specialists and other associated personnel must study a variety of books, articles, and reports on each of the topics in question. This process is time-consuming and rather difficult to follow because of the specialized nature of the material involved.

This book is an attempt to satisfy the pressing need for a single volume that considers maintainability, maintenance, and reliability topics. The material covered is treated in such a manner that the reader needs no previous related knowledge to understand it. The sources of most of the material presented are given in the reference section at the end of each chapter. At appropriate places, the book contains examples along with their solutions, and at the end of each chapter are numerous problems for testing reader comprehension.

The book is composed of 16 chapters. Chapter 1 presents the need for maintainability, maintenance, and reliability; the historical aspects of maintainability, maintenance, and reliability; important terms and definitions; and useful sources for obtaining information on maintainability, maintenance, and reliability. Chapter 2 reviews mathematical concepts considered useful to understand subsequent chapters. It covers topics such as Boolean algebra laws, probability properties, useful mathematical definitions, and probability distributions. Chapter 3 presents various introductory aspects of reliability.

Chapter 4 presents a number of reliability evaluation methods including fault tree analysis, the network reduction method, the delta–star method, the Markov method, and the supplementary variables method. Useful aspects of reliability management are presented in Chapter 5. Chapter 6 is devoted to mechanical and human reliability. It covers topics such as general mechanical failure causes and modes, safety factors, stress–strength interference theory modeling, human-error causes and classifications, human stress–performance effectiveness and stress factors, and methods for performing human-reliability analysis.

Chapter 7 presents introductory aspects of maintainability, including the need for maintainability, maintainability versus reliability and maintenance, and maintainability functions. Chapter 8 presents two important topics of maintainability: maintainability tools and specific maintainability design considerations. Chapter 9 is devoted to maintainability management and costing. It covers topics such as maintainability management tasks during the product life cycle, maintainability program plans, maintainability organization functions, maintainability design reviews, life cycle costing, maintainability investment cost elements, and life cycle cost estimation models.

Important aspects of human factors in maintainability are covered in Chapter 10. Chapter 11 covers topics such as the need for maintenance, maintenance engineering objectives, maintenance measures, safety in maintenance, and quality in maintenance. Chapter 12 is devoted to corrective and preventive maintenance. It covers many useful aspects of both corrective and preventive maintenance. Chapter 13 presents reliability-centered maintenance (RCM) topics, such as RCM goals and principles, RCM process, key RCM program elements, and RCM program measures.

Chapter 14 is devoted to maintenance management and costing, and Chapter 15 presents important aspects of human error in engineering maintenance. Chapter 16 covers three important topics of engineering maintenance: software maintenance, robotic maintenance, and medical equipment maintenance.

This book will be useful to many people including maintainability engineers; maintenance engineers; reliability specialists; design engineers; system engineers; engineering managers; graduate and senior undergraduate students of engineering; instructors and researchers of maintainability, maintenance, and reliability; and engineers-at-large.

I am deeply indebted to many people including students, friends, and colleagues for their input and encouragement at the moment of need. I thank my children Jasmine and Mark for their patience and intermittent disturbances that resulted in many desirable breaks. Last, but not least, I thank my other half, friend, and wife, Rosy, for her patience and for her help in proofreading.

B.S. Dhillon
Ottawa, Ontario

Author

B.S. Dhillon is a professor of mechanical engineering at the University of Ottawa. He has served as a chairman and director of the Mechanical Engineering Department/Engineering Management Program for over 10 years at the same institution. He has published over 320 articles and 29 books with publishers, including John Wiley & Sons (1981), Van Nostrand (1983), Marcel Dekker (1984), and Pergamon (1986). His books are used in over 70 countries and many of them have been translated into languages such as German, Russian, and Chinese. He is or has been on the editorial boards of nine international scientific journals. He has served as general chairman of two international conferences on reliability and quality control held in Los Angeles and Paris in 1987.

Professor Dhillon has served as a consultant to various organizations and has many years of experience in the industrial sector. At the University of Ottawa, he has taught engineering management, reliability, maintainability, safety, and related areas for over 25 years and he has also lectured in over 50 countries, including keynote addresses in North America, Europe, Asia, and Africa. In March 2004, he was a distinguished speaker at the conference and workshop on surgical errors sponsored by the White House Health and Safety Committee and the Pentagon, held at the Capitol Hill.

Professor Dhillon attended the University of Wales, where he earned a B.S. in electrical and electronic engineering and an M.S. in mechanical engineering. He was granted a Ph.D. in industrial engineering from the University of Windsor.

Contents

1 Introduction

1.1 NEED FOR MAINTAINABILITY, MAINTENANCE, AND RELIABILITY

Maintainability is becoming increasingly important because of the alarmingly high operating and support costs of systems and equipment. For example, each year the U.S. industry spends over $300 billion on plant maintenance and operations, and for the fiscal year 1997, the operation and maintenance budget request of the U.S. Department of Defense was $79 billion [1,2]. Thus, some of the objectives for applying maintainability engineering principles to systems and equipment are to reduce projected maintenance cost and time through design modifications directed at maintenance simplifications, to use maintainability data for estimating equipment availability or unavailability, and to determine labor hours and other related resources required to perform the projected maintenance.

Since the Industrial Revolution, maintenance of engineering systems has been a continuous challenge. Although impressive progress has been made in maintaining equipment in the field, maintenance of equipment is still a challenging issue because of various factors including complexity, cost, and competition. Each year billions of dollars are spent on engineering equipment maintenance worldwide, and it means there is a definite need for effective asset management and maintenance practices that can positively influence success factors such as quality, safety, price, speed of innovation, reliable delivery, and profitability.

The reliability of engineering systems has become an important issue during the design process because of the increasing dependence of our daily lives and schedules on the satisfactory functioning of these systems. Some examples of these systems are computers, aircraft, space satellites, nuclear power-generating reactors, automobiles, and trains. Some of the specific factors that play, directly or indirectly, an instrumental role in increasing the importance of reliability in designed systems include high acquisition cost; complexity; safety-, reliability-, and quality-related lawsuits; public pressures; and global competition.

These factors clearly indicate a definite need for maintainability, maintenance, and reliability professionals to work closely during the product design and operation phases. To achieve this goal successfully, it is absolutely essential that they have some understanding of each other's discipline. Once this goal is achieved, many of these professionals' work-related difficulties will be reduced to a tolerable level or disappear altogether, thus resulting in more reliable and maintainable or maintained systems.

1.2 HISTORY

This section presents an overview of historical developments in maintainability, maintenance, and reliability.

1.2.1 MAINTAINABILITY

An early reference to maintainability may be traced back to 1901 to the Army Signal Corps contract for development of the Wright brothers' airplane, in which it was clearly stated that the aircraft should be "simple to operate and maintain" [3]. In the modern context, the beginning of the maintainability discipline may be traced back to the period between World War II and the 1950s, when various efforts directly or indirectly concerned with maintainability were initiated. One example of these efforts is a 12-part series of articles that appeared in *Machine Design* in 1956 and covered topics such as design of electronic equipment for maintainability, recommendations for designing maintenance access in electronic equipment, and designing for installation [4].

In 1960 the U.S. Air Force (USAF) initiated a program for developing an effective systems approach to maintainability that ultimately resulted in the development of maintainability specification MIL-M-26512. Many other military documents concerning maintainability appeared in the latter part of the 1960s. Two examples of these documents are MIL-STD-470 [5] and MIL-HDBK-472 [6].

The first commercially available book on maintainability, *Electronic Maintainability* appeared in 1960 [7]. Over the years many other developments in the maintainability field have taken place, and a detailed history of maintainability engineering is available in References 4 and 8.

1.2.2 MAINTENANCE

Although humans have felt the need to maintain their equipment since the beginning of time, the beginning of modern engineering maintenance may be regarded as the development of the steam engine by James Watt (1736–1819) in 1769 in Great Britain [9]. In the United States the magazine *Factory* first appeared in 1882 and has played a pivotal role in the development of the maintenance field [10]. In 1886 a book on maintenance of railways was published [11].

In the 1950s the term *preventive maintenance* was coined, and in 1957 a handbook on maintenance engineering was published [12]. Over the years many other developments in the field of engineering maintenance have taken place, and today many universities and other institutions offer academic programs on the subject.

1.2.3 RELIABILITY

The history of reliability engineering may be traced back to World War II, when the Germans are reported to have first introduced the reliability concept to improve the reliability of their V1 and V2 rockets. In 1950 the U.S. Department of Defense established an ad hoc committee on reliability, and in 1952 it was transformed to a permanent group called the Advisory Committee on the Reliability of Electronic Equipment (AGREE) [13]. The committee released its report in 1957 [14].

In 1954 a national symposium on reliability was held for the first time in the United States, and in 1957 the USAF released the first military specification (MIL-R-25717 [USAF]), "Reliability Assurance Program for Electronic Equipment" [15]. In 1962 the first master's degree program in system reliability engineering was started at the Air Force Institute of Technology in Dayton, Ohio.

Over the years many other developments in reliability engineering have occurred, and a detailed history of reliability engineering is available in Reference 14.

1.3 MAINTAINABILITY, MAINTENANCE, AND RELIABILITY TERMS AND DEFINITIONS

Many terms and definitions are used in maintainability, maintenance, and reliability engineering work. The section presents some of the frequently used terms and definitions in these three areas taken from various sources [3,16–22]:

Maintainability: The probability that a failed item will be restored to its satisfactory operational state

Maintenance: All actions necessary for retaining an item or equipment in, or restoring it to, a specified condition

Reliability: The probability that an item will perform its assigned mission satisfactorily for the stated time period when used according to the specified conditions

Availability: The probability that an item is available for use when required

Mission time: The time during which the item is carrying out its assigned mission

Downtime: The total time during which the item is not in satisfactory operating state

Logistic time: The portion of downtime occupied by the wait for a required part or tool

Failure: The inability of an item to operate within the defined guidelines

Serviceability: The degree of ease or difficulty with which an item can be restored to its working condition

Redundancy: The existence of more than one means for accomplishing a stated function

Failure mode: The abnormality of an item's performance that causes the item to be considered to have failed

Human reliability: The probability of accomplishing a task successfully by humans at any required stage in the system operation with a given minimum time limit (if the time requirement is stated)

Useful life: The length of time a product operates within a tolerable level of failure rate

Maintenance concept: A statement of the overall concept of the product specification or policy that controls the type of maintenance action to be taken for the product under consideration.

Corrective maintenance: The repair or unscheduled maintenance to return items or equipment to a specified state, performed because maintenance personnel or others perceived deficiencies or failures

Continuous task: A task that involves some kind of tracking activity (e.g., monitoring a changing situation)

Human performance: A measure of human functions and actions under some specified conditions

Active redundancy: A type of redundancy in which all redundant units are functioning simultaneously

Human error: The failure to carry out a specified task (or the performance of a forbidden action) that could result in disruption of scheduled operations or damage to property or equipment

Active repair time: The period of downtime when repair personnel are active to effect a repair

Inspection: The qualitative observation of an item's condition or performance

Overhaul: A comprehensive inspection and restoration of a piece of equipment or an item to an acceptable level at a durability time or usage limit

1.4 USEFUL INFORMATION ON MAINTAINABILITY, MAINTENANCE, AND RELIABILITY

There are many different sources for obtaining maintainability-, maintenance-, and reliability-related information. This section lists some of the most useful sources for obtaining such information.

1.4.1 JOURNALS AND MAGAZINES

- *IEEE Transactions on Reliability*
- *Journal of Quality in Maintenance Engineering*
- *Reliability Review*
- *Maintenance and Asset Management Journal*
- *Maintenance Technology*
- *Reliability Engineering and System Safety*
- *Industrial Maintenance and Plant Operation*
- *International Journal of Reliability, Quality, and Safety Engineering*
- *Maintenance Journal*
- *Quality and Reliability Engineering International*
- *Quality and Reliability Management*
- *Reliability: The Magazine for Improved Plant Reliability*
- *RAMS ASIA (Reliability, Availability, Maintainability, and Safety (RAMS) Quarterly Journal)*
- *Journal of Software Maintenance and Evolution: Research and Practice*
- *PEM (Plant Engineering and Maintenance)*

1.4.2 Books

- Blanchard, B.S., Verma, D., and Peterson, E.L., *Maintainability: A Key to Effective Serviceability and Maintenance Management*, John Wiley & Sons, New York, 1995
- Goldman, A.S. and Slattery, T.B., *Maintainability*, John Wiley & Sons, New York, 1964
- Higgins, L.R., Mobley, R.K., and Smith, R., Eds., *Maintenance Engineering Handbook*, McGraw-Hill, New York, 2002
- August, J., *RCM Guidebook*, Penn Well, Tulsa, OK, 2004
- Dhillon, B.S., *Design Reliability: Fundamentals and Applications*, CRC Press, Boca Raton, FL, 1999
- Blank, R., *The Basics of Reliability*, Productivity Press, New York, 2004
- *Maintainability Toolkit*, Reliability Analysis Center, Griffis Air Force Base, Rome, NY, 2000
- Cunningham, C.E. and Cox, W., *Applied Maintainability Engineering*, John Wiley & Sons, New York, 1972
- Dhillon, B.S., *Engineering Maintainability*, Gulf Publishing Company, Houston, TX, 1999
- Gertsbakh, I.B., *Reliability Theory: With Applications to Preventive Maintenance*, Springer, New York, 2001
- Shooman, M.L., *Probabilistic Reliability: An Engineering Approach*, McGraw-Hill, New York, 1968
- Evans, J.W. and Evan, J.Y., *Product Integrity and Reliability in Design*, Springer, New York, 2001
- Blanchard, B.S. and Lowery, E.E., *Maintainability Principles and Practices*, McGraw-Hill, New York, 1969
- Dhillon, B.S., *Engineering Maintenance: A Modern Approach*, CRC Press, Boca Raton, FL, 2002
- Elsayed, E.A., *Reliability Engineering*, Addison-Wesley, Reading, MA, 1996
- Grant Ireson, W., Coombs, C.F., and Moss, R.Y., Eds., *Handbook of Reliability Engineering and Management*, McGraw-Hill, New York, 1996
- Niebel, B.W., *Engineering Maintenance Management*, Marcel Dekker, New York, 1994
- Moubray, J., *Reliability-Centered Maintenance*, Industrial Press, New York, 1997
- Smith, D.J. and Babb, R.H., *Maintainability Engineering*, Pitman, New York, 1973
- Kelly, A., *Maintenance Strategy*, Butterworth-Heinemann, Oxford, UK, 1997

1.4.3 Data Information Sources

- Government Industry Data Exchange Program (GIDEP) Operations Center, Department of the Navy, Corona, CA
- IEC 706 PT3, *Guide on Maintainability of Equipment*, Part III: Sections Six and Seven, *Verification and Collection, Analysis and Presentation of Data*, 1st ed., International Electro-Technical Commission, Geneva, Switzerland

- National Technical Information Service (NTIS), 5285 Port Royal Road, Springfield, VA
- RACEEMDI, Electronics Equipment Maintainability Data, Reliability Analysis Center, Rome Air Development Center, Griffis Air Force Base, Rome, NY
- Defense Technical Information Center, DTIC-FDAC, 8725 John J. Kingman Road, Suite 0944, Fort Belvoir, VA
- American National Standards Institute (ANSI), 11 West 42nd Street, New York, NY 10036
- Space Documentation Service, European Space Agency, Via Galileo Galilei, Frascati 00044, Italy
- National Aeronautics and Space Administration (NASA) Parts Reliability Information Center, George C. Marshall Space Flight Center, Huntsville, AL
- System Reliability Service, Safety and Reliability Directorate, UKAEA, Wigshaw Lane, Culcheth, Warrington, U.K.

1.4.4 ORGANIZATIONS

- Society of Logistics Engineers, 8100 Professional Place, Suite 211, Hyattsville, MD
- Reliability Society, IEEE, P.O. Box 1331, Piscataway, NJ
- Society for Maintenance and Reliability Professionals, 401 N. Michigan Avenue, Chicago, IL
- American Society for Quality, Reliability Division, 600 North Plankinton Avenue, Milwaukee, WI
- American Institute of Plant Engineers, 539 South Lexington Place, Anaheim, CA
- Japan Institute of Plant Maintenance, Shuwa Shiba-koen3-Chome Building, 3-1-38, Shiba-Koen, Minato-Ku, Tokyo, Japan
- Society of Reliability Engineers (SRE), e-mail address: webmaster@sre.org
- Society for Machinery Failure Prevention Technology, 4193 Sudley Road, Haymarket, VA
- The Institution of Plant Engineers, 77 Great Peter Street, London, UK
- System Safety Society, 14252 Culver Drive, Suite A-261, Irvine, CA
- Maintenance Engineering Society of Australia (MESA), 11 National Circuit, Barton, ACT, Australia

1.5 PROBLEMS

1. Define the following terms:
   - Maintainability
   - Maintenance
   - Reliability
2. List at least three useful sources for obtaining maintainability, maintenance, and reliability-related information.

3. Discuss the need for maintainability, maintenance, and reliability.
4. Discuss the history of maintainability.
5. Compare maintainability with maintenance.
6. Define the following two terms:
   - Failure mode
   - Human reliability
7. What is the difference between reliability and maintainability?
8. What is the difference between logistic time and downtime?
9. List three most useful journals for obtaining maintainability-related information.
10. Compare historical developments in reliability and maintainability fields.

REFERENCES

1. Latino, C.J., *Hidden Treasure: Eliminating Chronic Failures Can Cut Maintenance Costs Up to 60%*, Report, Reliability Center, Hopewell, VA, 1999.
2. 1997 *DOD Budget: Potential Reductions to Operation and Maintenance Program*, U.S. General Accounting Office, Washington, DC, 1996.
3. *Engineering Design Handbook: Maintainability Engineering Theory and Practice*, AMCP 706-133, Department of Defense, Washington, DC, 1976.
4. Retterer, B.L. and Kowalski, R.A., Maintainability: a historical perspective, *IEEE Transactions on Reliability*, 33, 56–61, 1984.
5. *Maintainability Program Requirements*, MIL-STD-470, Department of Defense, Washington, DC, 1966.
6. *Maintainability Prediction*, MIL-HDBK-472, Department of Defense, Washington, DC, 1966.
7. Akenbrandt, F.L., Ed., *Electronic Maintainability*, Engineering Publishers, Elizabeth, NJ, 1960.
8. Dhillon, B.S., *Engineering Maintainability*, Gulf Publishing Company, Houston, TX, 1999.
9. *The Volume Library: A Modern Authoritative Reference for Home and School Use*, The South-Western Company, Nashville, TN, 1993.
10. *Factory*, McGraw-Hill, New York, 1882–1968.
11. Kirkman, M.M., *Maintenance of Railways*, C.N. Trivess Printers, Chicago, 1886.
12. Morrow, L.C., Ed., *Maintenance Engineering Handbook*, McGraw-Hill, New York, 1957.
13. Advisory Group on Reliability of Electronic Equipment (AGREE), *Reliability of Military Electronic Equipment*, Office of the Assistant Secretary of Defense (Research and Engineering), Department of Defense, Washington, DC, 1957.
14. Coppola, A., Reliability engineering of electronic equipment: a historical perspective, *IEEE Transactions on Reliability*, 33, 29–35, 1984.
15. *Reliability Assurance Program for Electronic Equipment*, MIL-R-25717 (USAF), Department of Defense, Washington, DC, 1957.
16. Von Alven, W.H., Ed., *Reliability Engineering*, Prentice Hall, Englewood Cliffs, NJ, 1964.
17. Naresky, J..J., Reliability definitions, *IEEE Transactions on Reliability*, 19, 198–200, 1970.
18. *Definitions of Terms for Reliability and Maintainability*, MIL-STD-721C, Department of Defense, Washington, DC, 1981.

19. *Policies Governing Maintenance Engineering within the Department of Defense*, DOD INST. 4151.12, Department of Defense, Washington, DC, 1968.
20. Mckenna, T. and Oliverson, R., *Glossary of Reliability and Maintenance Terms*, Gulf Publishing Company, Houston, TX, 1997.
21. Omdahl, T.P., Ed., *Reliability, Availability, and Maintainability (RAM) Dictionary*, ASQC Quality Press, Milwaukee, WI, 1988.
22. *Engineering Design Handbook: Maintenance Engineering Techniques*, AMCP 706-132, Department of the Army, Washington, DC, 1975.

2 Maintainability, Maintenance, and Reliability Mathematics

2.1 INTRODUCTION

Just like in other areas of engineering, mathematics plays a pivotal role in the fields of maintainability, maintenance, and reliability engineering. However, its applications in engineering in general are relatively new.

Although the origin of our current number symbols goes back to 250 B.C., the history of probability, as probability plays a central role in the analysis of maintainability, maintenance, and reliability problems, may only be traced back to the sixteenth-century writings of Girolamo Cardano (1501–1576) [1], who wrote a gambler's manual and considered some interesting questions on probability. In the seventeenth century Pierre Fermat (1601–1665) and Blaise Pascal (1623–1662) solved the problem of dividing the winnings in a game of chance correctly and independently. In the eighteenth century, Pierre Laplace (1749–1827) and Karl Gauss (1777–1855) further developed probability concepts and successfully applied them to areas other than games of chance [2].

A detailed history of probability and other areas of mathematics is available in Reference 1. This chapter presents various aspects of mathematics that will be useful in understanding subsequent chapters of this volume.

2.2 BOOLEAN ALGEBRA LAWS AND PROBABILITY PROPERTIES

Boolean algebra plays an important role in probability theory and is named after its originator, the mathematician George Boole (1813–1864) [3]. Some of its laws are as follows [3,4]:

- Idempotent Law:

$$A + A = A \qquad (2.1)$$

$$A \cdot A = A \qquad (2.2)$$

where A is an arbitrary set or event, + denotes the union of sets or events, and dot (·) denotes the intersection of sets or events. Sometimes Equation 2.2 or others are written without the dot, but they still convey the same meaning.

- Associative Law:

$$(AB)\ C = A\ (BC) \tag{2.3}$$

where B is an arbitrary set or event, and C is an arbitrary set or event.

$$(A + B) + C = A + (B + C) \tag{2.4}$$

- Absorption Law:

$$A + (AB) = A \tag{2.5}$$

$$A\ (A + B) = A \tag{2.6}$$

- Distributive Law:

$$A\ (B + C) = AB + AC \tag{2.7}$$

$$A + BC = (A + B)\ (A + C) \tag{2.8}$$

- Commutative Law:

$$A + B = B + A \tag{2.9}$$

$$AB = BA \tag{2.10}$$

There are many properties of probability. Some of these are as follows [4–6]:

- The probability of occurrence of event, for example, X, is always

$$0 \le P(X) \le 1 \tag{2.11}$$

where $P(X)$ is the probability of occurrence of event X.

- The probability of the sample space S is

$$P(S) = 1 \tag{2.12}$$

- The probability of the negation of the sample space S is given by

$$P(\bar{S}) = 0 \tag{2.13}$$

where $\bar{S}$ is the negation of the sample space S.

- The probability of occurrence and nonoccurrence of an event, for example, X, is given by

$$P(X) + P(\bar{X}) = 1 \tag{2.14}$$

where $\bar{X}$ is the negation of event X and $P(\bar{X})$ is the probability of nonoccurrence of event X.

- The probability of an intersection of m independent events is given by

$$P(X_1 X_2 X_3 \ldots X_m) = P(X_1) P(X_2) P(X_3) \ldots P(X_m) \tag{2.15}$$

where X_i is the ith event for $i = 1, 2, 3, \ldots, m$. $P(X_i)$ is the probability of occurrence of event X_i for $i = 1, 2, 3, \ldots, m$.

- The probability of the union of m mutually exclusive events is expressed by

$$P(X_1 + X_2 + X_3 + \ldots + X_m) = \sum_{i=1}^{m} P(X_i) \tag{2.16}$$

- The probability of the union of m independent events is

$$P(X_1 + X_2 + X_3 + \ldots + X_m) = 1 - \prod_{i=1}^{m} \left(1 - P(X_i)\right) \tag{2.17}$$

It is to be noted that for very small values of $P(X_i)$; for $i = 1, 2, 3, \ldots, m$, Equation 2.16 and Equation 2.17 yield almost the same result. This is demonstrated through the following example.

Example 2.1
Assume that in Equation 2.16 and Equation 2.17 we have $m = 2$, $P(X_1) = 0.01$, and $P(X_2) = 0.08$. Calculate the probability of the union of events X_1 and X_2 using these two equations and comment on the resulting values.
Using the given data values in Equation 2.16 yields

$$\begin{aligned} P(X_1 + X_2) &= P(X_1) + P(X_2) \\ &= 0.01 + 0.08 \\ &= 0.09 \end{aligned}$$

By substituting the same data values into Equation 2.17, we get

$$P(X_1+X_2)=P(X_1)+P(X_2)-P(X_1)\,P(X_2)$$
$$=0.01+0.08-(0.01)(0.08)$$
$$=0.0892$$

The above two calculated values are very close.

2.3 USEFUL DEFINITIONS AND PROBABILITY DISTRIBUTIONS

Many mathematical definitions and probability distributions are used to perform various types of maintainability, maintenance, and reliability studies. This section presents some of those considered to be quite useful for such purposes.

2.3.1 PROBABILITY

This is defined by [5]

$$P(A)=\lim_{n\to\infty}\left(\frac{N}{n}\right) \tag{2.18}$$

where $P(A)$ is the probability of occurrence of event A and N is the total number of times that event A occurs in the n repeated experiments.

2.3.2 CUMULATIVE DISTRIBUTION FUNCTION

For continuous random variables, the cumulative distribution function is expressed by [5]

$$F(t)=\int_{-\infty}^{t} f(y)\,dy \tag{2.19}$$

where t is time, $f(y)$ is the probability density function, and $F(t)$ is the cumulative distribution function.

For $t = \infty$, Equation 2.19 yields

$$F(\infty)=\int_{-\infty}^{\infty} f(y)\,dy \tag{2.20}$$
$$=1$$

This means that the total area under the probability density curve is equal to unity.

2.3.3 PROBABILITY DENSITY FUNCTION

This is defined by [5]

$$\frac{dF(t)}{dt} = \frac{d\left[\int_{-\infty}^{t} f(y)\,dy\right]}{dt} \tag{2.21}$$

$$= f(t)$$

2.3.4 RELIABILITY FUNCTION

This is defined by

$$R(t) = 1 - F(t)$$

$$= 1 - \int_{-\infty}^{t} f(y)\,dy \tag{2.22}$$

where $f(y)$ is the failure density function and $R(t)$ is the reliability function.

2.3.5 CONTINUOUS RANDOM VARIABLE EXPECTED VALUE

The expected value of a continuous random variable is defined by [6]

$$E(t) = \int_{-\infty}^{\infty} t\,f(t)\,dt \tag{2.23}$$

where $E(t)$ is the expected value or mean of a continuous random variable, t.

2.3.6 DISCRETE RANDOM VARIABLE EXPECTED VALUE

The expected value of a discrete random variable is expressed by [5,6]

$$E(t) = \sum_{j=1}^{n} t_j\,f(t_j) \tag{2.24}$$

where $E(t)$ is the expected value of a discrete random variable t and n is the number of discrete values of the random variable t.

2.3.7 EXPONENTIAL DISTRIBUTION

This is a widely used probability distribution in maintainability, maintenance, and reliability work. Two basic reasons for its widespread use are that it is easy to handle in performing various types of analyses and the constant failure rate of many engineering items during their useful lives, particularly the electronic ones [7].

The distribution probability density function is defined by

$$f(t) = \lambda e^{-\lambda t} \quad for \quad t \geq 0, \ \lambda > 0 \tag{2.25}$$

where t is time, $f(t)$ is the probability density function, and λ is the distribution parameter, which in reliability work is known as the constant failure rate.

By substituting Equation 2.25 into Equation 2.19, we get

$$F(t) = \int_0^t \lambda e^{-\lambda x} \, dx \tag{2.26}$$
$$= 1 - e^{-\lambda t}$$

Example 2.2
The failure rate of an electronic item is 0.005 failures per hour, and its times to failure are defined by the following function:

$$f(t) = 0.005 \ e^{-(0.005) \, t}, \quad for \quad t \geq 0 \tag{2.27}$$

where
t is time and $f(t)$ is the failure density function.

Calculate the probability that the item will fail during a 50-hour mission.
Using the given data, substituting Equation 2.27 into Equation 2.19 yields

$$F(50) = \int_0^{50} 0.005 \, e^{-(0.005) \, x} \, dx$$
$$= 1 - e^{-(0.005)(50)}$$
$$= 0.2211$$

This means there is an approximately 22% chance that the item will fail during the specified time period.

2.3.8 RAYLEIGH DISTRIBUTION

This distribution is known after its originator, John Rayleigh (1842–1919), and is sometime used in reliability-related studies [1]. Its probability density function is defined by

$$f(t) = \left(\frac{2}{\alpha^2} \right) t \, e^{-\left(\frac{t}{\alpha} \right)^2}, \quad for \quad t \geq 0, \ \alpha > 0 \tag{2.28}$$

where α is the distribution parameter.

By substituting Equation 2.28 into Equation 2.19, we get

$$F(t) = \int_0^t \left(\frac{2}{\alpha^2}\right) x \, e^{-\left(\frac{x}{\alpha}\right)^2} dx$$

$$= 1 - e^{-\left(\frac{t}{\alpha}\right)^2}$$

(2.29)

2.3.9 WEIBULL DISTRIBUTION

This distribution was developed by W. Weibull in the early 1950s and it can be used to represent many different physical phenomena [8]. Its probability density function is expressed by

$$f(t) = \frac{\beta}{\alpha^\beta} t^{\beta-1} e^{-\left(\frac{t}{\alpha}\right)^\beta}, \quad for \quad t \geq 0, \, \alpha > 0, \beta > 0$$

(2.30)

where α and β are the distribution scale and shape parameters, respectively.

Substituting Equation 2.30 into Equation 2.19 yields the following equation for the cumulative distribution function:

$$F(t) = \int_0^t \frac{\beta}{\alpha^\beta} x^{\beta-1} e^{-\left(\frac{x}{\alpha}\right)^\beta} dx$$

$$= 1 - e^{-\left(\frac{t}{\alpha}\right)^\beta}$$

(2.31)

Both exponential and Rayleigh distributions are the special cases of Weibull distribution for $\beta = 1$ and 2, respectively.

2.3.10 GAMMA DISTRIBUTION

This is a two-parameter distribution, and its probability density function is defined by [9]

$$f(t) = \frac{\lambda(\lambda t)^{k-1}}{\Gamma(k)} e^{-\lambda t}, \quad for \quad t \geq 0, \, \lambda > 0, k > 0$$

(2.32)

where k is the distribution shape parameter, $\Gamma(k)$ is the gamma function, and $\lambda = \frac{1}{\alpha}$ (α is the distribution scale parameter).

By substituting Equation 2.32 into Equation 2.19 we get

$$F(t) = \int_0^t \frac{\lambda(\lambda x)^{k-1}}{\Gamma(k)} e^{-\lambda x} dx$$

$$= 1 - \frac{\Gamma(k, \lambda t)}{\Gamma(k)}$$

(2.33)

where $\Gamma(k, \lambda t)$ is the incomplete gamma function.

For $k = 1$, integer values of k, $\lambda = 0.5$, and $k = 0.5n$ (where n is the number of degrees of freedom), the gamma distribution becomes the exponential distribution, the special Erlangian distribution, and the chi-square distribution, respectively [10].

2.3.11 NORMAL DISTRIBUTION

This is a widely used probability distribution and is also known as the Gaussian distribution after Carl Friedrich Gauss (1777–1855). Its probability density function is defined by

$$f(t) = \frac{1}{\sigma\sqrt{2\Pi}} \exp\left[-\frac{(t-\mu)^2}{2\sigma^2}\right], \ \ for \ \ \infty < t < +\infty \tag{2.34}$$

where μ is the mean, and σ is the standard deviation.

Substituting Equation 2.34 into Equation 2.19 yields the following equation for the cumulative distribution function:

$$f(t) = \int_{-\infty}^{t} \frac{1}{\sigma\sqrt{2\Pi}} \exp\left[-\frac{(t-\mu)^2}{2\sigma^2}\right] dt$$

$$= \frac{1}{\sigma\sqrt{2\Pi}} \int_{-\infty}^{t} \exp\left[-\frac{(x-\mu)^2}{2\sigma^2}\right] dx \tag{2.35}$$

2.3.12 LOGNORMAL DISTRIBUTION

This distribution is quite useful to represent the distribution of failed equipment repair times. The probability density function of the lognormal distribution is defined by

$$f(t) = \frac{1}{t\alpha\sqrt{2\Pi}} \exp\left[-\frac{(\ln t - \mu)^2}{2\alpha^2}\right], for \ \ t \geq 0 \tag{2.36}$$

where α and μ are the distribution parameters.

By substituting Equation 2.36 into Equation 2.19, we get the following equation for the cumulative distribution function:

$$F(t) = \frac{1}{\alpha\sqrt{2\Pi}} \int_{0}^{t} \frac{1}{x} \exp\left[-\frac{(\ln x - \mu)^2}{2\alpha^2}\right] dx \tag{2.37}$$

2.3.13 BINOMIAL DISTRIBUTION

This discrete random variable distribution has applications in many combinational-type reliability problems and is also known as the Bernoulli distribution after Jakob Bernoulli (1654–1705). The binomial probability density function is expressed by

$$f(x) = \binom{n}{x} p^x q^{n-x}, \; for \; x = 0, 1, 2, ..., n \tag{2.38}$$

$$where \; \binom{n}{x} = \frac{n!}{x!(n-x)!}$$

x is the total number of failures in n trials, q is the single-trial failure probability, and p is the single-trial success probability.

The sum of p and q is always equal to unity. The cumulative distribution function is given by

$$F(x) = \sum_{j=0}^{x} \binom{n}{j} p^j q^{n-j} \tag{2.39}$$

where $F(x)$ is the cumulative distribution function or the probability of x or less failures in n trials.

2.4 LAPLACE TRANSFORMS AND THEIR APPLICATIONS TO DIFFERENTIAL EQUATIONS

Laplace transforms are used in performing various types of maintainability, maintenance, and reliability studies. The Laplace transform of the function $f(t)$ is defined by

$$f(s) = \int_{0}^{\infty} f(t) e^{-st} \, dt \tag{2.40}$$

where t is the time variable, s is the Laplace transform variable, and $f(s)$ is the Laplace transform of $f(t)$.

Example 2.3
Obtain the Laplace transform of the following function:

$$f(t) = e^{-\lambda t} \tag{2.41}$$

where t is time and λ is a constant.

Substituting Equation 2.41 into Equation 2.40 yields

$$
\begin{aligned}
f(s) &= \int_0^\infty e^{-\lambda t} e^{-st}\, dt \\
&= \frac{e^{-(s+\lambda)t}}{-(s+\lambda)} \Big|_0^\infty \\
&= \frac{1}{s+\lambda}
\end{aligned}
\tag{2.42}
$$

Laplace transforms of some commonly occurring functions in performing various types of mathematical maintainability, maintenance, and reliability studies are presented in Table 2.1. Laplace transforms of a wide range of mathematical functions are available in References [11–13].

2.4.1 LAPLACE TRANSFORMS: INITIAL AND FINAL VALUE THEOREMS

The Laplace transform of the initial value theorem is expressed by

$$
\lim_{t \to 0} f(t) = \lim_{s \to \infty} s f(s)
\tag{2.43}
$$

If the following limits exist, then the final-value theorem may be expressed as

$$
\lim_{t \to \infty} f(t) = \lim_{s \to 0} s f(s)
\tag{2.44}
$$

TABLE 2.1
Laplace Transforms of Some Frequently Occurring Functions in Maintainability, Maintenance, and Reliability Studies

Number	$f(t)$	$f(s)$
1	c (a constant)	$\dfrac{c}{s}$
2	e^{-at}	$\dfrac{1}{s+a}$
3	$t\,e^{-at}$	$\dfrac{1}{(s+a)^2}$
4	$\dfrac{df(t)}{dt}$	$sf(s) - f(0)$
5	t^K, for $K = 0, 1, 2, 3, \ldots.$	$\dfrac{K!}{s^{K+1}}$

Example 2.4

Prove by using the following equation that the results obtained using the left-hand side and the right-hand side of Equation 2.44 are the same:

$$f(t) = \frac{\mu}{\lambda + \mu} + \frac{\lambda}{\lambda + \mu} e^{-(\lambda + \mu)t} \tag{2.45}$$

where λ and μ are constants.

Substituting Equation 2.45 into the left-hand side of Equation 2.44 yields

$$\lim_{t \to \infty} \left[\frac{\mu}{\lambda + \mu} + \frac{\lambda}{\lambda + \mu} e^{-(\lambda + \mu)t} \right] = \frac{\mu}{\lambda + \mu} \tag{2.46}$$

Substituting Equation 2.45 into Equation 2.40 yields

$$f(s) = \frac{(s + \mu)}{s(s + \mu + \lambda)} \tag{2.47}$$

By substituting Equation 2.47 into the right-hand side of Equation 2.44 we get

$$\lim_{s \to 0} \left[\frac{s(s + \mu)}{s(s + \mu + \lambda)} \right] = \frac{\mu}{\lambda + \mu} \tag{2.48}$$

Equation 2.46 and Equation 2.48 are identical, proving that the results obtained using both the left-hand side and the right-hand side of Equation 2.44 are the same.

2.4.2 LAPLACE TRANSFORM APPLICATION TO DIFFERENTIAL EQUATIONS

Sometimes maintainability, maintenance, and reliability studies may require solutions to a system of linear first-order differential equations. Under such circumstances, Laplace transforms are a very useful tool. Their application to finding solutions to a set of differential equations is demonstrated through the following example.

Example 2.5

An engineering system is described by the following two differential equations:

$$\frac{dP_n(t)}{dt} + \theta P_n(t) = 0 \tag{2.49}$$

$$\frac{dP_f(t)}{dt} - P_n(t)\theta = 0 \tag{2.50}$$

where θ is the system constant failure rate and $P_j(t)$ is the probability that the system is in state j at time t for $j = 0$ (operating normally) and $j = f$ (failed).

At time $t = 0$, $P_n(0) = 1$, and $P_f(0) = 0$.

Find solutions to Equation 2.49 and Equation 2.50 by using Laplace transforms. Taking Laplace transforms of Equation 2.49 and Equation 2.50 we get

$$sP_n(s) - P_n(0) + \theta P_n(s) = 0 \tag{2.51}$$

$$sP_f(s) - P_f(0) - \theta P_f(s) = 0 \tag{2.52}$$

Using the given initial conditions and then solving Equation 2.51 and Equation 2.52, we get

$$P_n(s) = \frac{1}{s+\theta} \tag{2.53}$$

$$P_f(s) = \frac{\theta}{s(s+\theta)} \tag{2.54}$$

Taking inverse Laplace transforms of Equations 2.53 and Equation 2.54 yields

$$P_n(t) = e^{-\theta t} \tag{2.55}$$

$$P_f(t) = 1 - e^{-\theta t} \tag{2.56}$$

Thus, Equation 2.55 and Equation 2.56 are the solutions to Equation 2.49 and Equation 2.50.

2.5 PROBLEMS

1. Write an essay on the history of probability.
2. What are the idempotent law and absorption law of Boolean algebra?
3. Write down at least five properties of probability.
4. Assume that in Equation 2.17 we have $m = 3$, $P(X_1) = 0.02$, $P(X_2) = 0.06$, and $P(X_3) = 0.08$. Calculate the probability of the union of events X_1, X_2, and X_3.
5. Write down the mathematical probability definition.
6. Prove that the total area under the probability density curve is equal to unity.

7. Write down probability density functions for the following distributions:
   - Weibull distribution
   - Lognormal distribution
8. Obtain the Laplace transform for the following function:

$$f(t) = t^n, \quad for \quad n = 0, 1, 2, 3, \ldots \tag{2.57}$$

9. Prove Equation 2.44.
10. Find the solution to the following first-order differential equation describing an engineering system for time $t = 0$ and $P_0(0) = 1$.

$$\frac{d P_0(t)}{dt} + \left(\sum_{j=1}^{K} \theta_i \right) P_0(t) = 0 \tag{2.58}$$

where K is the number of failure modes, θ_j is the constant failure rate of the system failing in failure mode j, and $P_0(t)$ is the probability of the system being operational at time t.

REFERENCES

1. Eves, H., *An Introduction to the History of Mathematics*, Holt, Reinhart, and Winston, New York, 1976.
2. Shooman, M.L., *Probabilistic Reliability: An Engineering Approach*, McGraw-Hill, New York, 1968.
3. Lipschutz, S., *Set Theory and Related Topics*, McGraw-Hill, New York, 1964.
4. NUREG-0492, *Fault Tree Handbook*, Nuclear Regulatory Commission, Washington, DC, 1981.
5. Mann, N.R., Schafer, R.E., and Singpurwalla, N.D., *Methods of Statistical Analysis of Reliability and Life Data*, John Wiley & Sons, New York, 1974.
6. Lipschutz, S., *Probability*, McGraw-Hill, New York, 1965.
7. Davis, D.J., An analysis of some failure data, *Journal of the American Statistical Association*, June, 113–150, 1952.
8. Weibull, W., A statistical distribution function of wide applicability, *Journal of Applied Mechanics*, 18, 293–297, 1951.
9. Gupta, S. and Groll, P., Gamma distribution in acceptance sampling based on life tests, *Journal of the American Statistical Association*, December, 942–970, 1961.
10. Dhillon, B.S., *Mechanical Reliability: Theory, Models, and Applications*, American Institute of Aeronautics and Astronautics, Washington, DC, 1988.
11. Oberhettinger, F. and Badic, L., *Tables of Laplace Transforms*, Springer-Verlag, New York, 1973.
12. Spiegel, M.R., *Laplace Transforms*, McGraw-Hill, New York, 1965.
13. Widder, D.V., *The Laplace Transform*, Princeton University Press, Princeton, NJ, 1941.

3 Introduction to Engineering Reliability

3.1 NEED FOR RELIABILITY

The reliability of engineering systems has become an important issue during their design because of the increasing dependence of our daily lives and schedules on the satisfactory functioning of these systems. Some examples of these systems are aircraft, trains, computers, automobiles, space satellites, and nuclear power–generating reactors. Many of these systems have become highly complex and sophisticated. For example, today a typical Boeing 747 jumbo airplane is made of approximately 4.5 million parts, including fasteners [1]. Most of these parts must function normally for the aircraft to fly successfully.

Normally, the required reliability of engineering systems is specified in the design specification, and during the design phase every effort is made to fulfill this requirement effectively. Some of the factors that play a key role in increasing the importance of reliability in designed systems are the increasing number of reliability- and quality-related lawsuits, competition, public pressures, high acquisition cost, past well-publicized system failures, loss of prestige, and complex and sophisticated systems.

This chapter presents various introductory aspects of engineering reliability.

3.2 BATHTUB HAZARD RATE CONCEPT

This is a well-known concept used to represent failure behavior of various engineering items because the failure rate of such items is a function of time (i.e., it changes with time). A bathtub hazard rate curve is shown in Figure 3.1. It is divided into three regions (i.e., Region I, Region II, and Region III). Region I is known as the burn-in region, debugging region, infant mortality region, or break-in region. During this period or region the item hazard rate (i.e., time-dependent failure rate) decreases because of failures occurring for reasons such as listed in Table 3.1 [2]. Region II is referred to as the "useful life period," during which the item hazard rate remains constant. Some of the reasons for the occurrence failure in this region are presented in Table 3.1. Region III is known as the "wear-out period," during which the hazard rate increases because of failures occurring for reasons such as presented in Table 3.1.

Mathematically, the bathtub hazard rate curve shown in Figure 3.1 can be represented by using the following function [3]:

$$\lambda(t) = \theta \lambda \beta t^{\beta-1} + (1-\theta) b t^{b-1} \alpha e^{\alpha t^b} \tag{3.1}$$

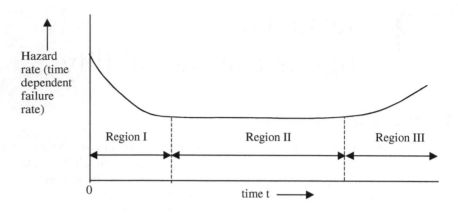

FIGURE 3.1 Bathtub hazard rate curve.

TABLE 3.1
Reasons for the Occurrence of Failures in the Three Regions of the Bathtub Hazard Rate Curve

Region	Reason
I: Burn-in period	Poor manufacturing methods
	Poor processes
	Poor quality control
	Poor debugging
	Human error
	Substandard materials and workmanship
II: Useful life period	Low safety factors
	Undetectable defects
	Human errors
	Abuse
	Higher random stress than expected
	Natural failures
III: Wear-out period	Wear caused by friction
	Poor maintenance
	Incorrect overhaul practices
	Corrosion and creep
	Short designed-in life of the item
	Wear caused by aging

for β, b, λ, and $\alpha > 0$; $0 \leq \theta \leq 1$; $\beta = 0.5$, $b = 1$, and $t \geq 0$ and where t is time, $\lambda(t)$ is the hazard rate or time t–dependent failure rate, α and λ are the scale parameters, and β and b are the shape parameters.

3.3 GENERAL RELIABILITY ANALYSIS FORMULAS

A number of formulas, based on the reliability function, frequently are used to perform various types of reliability analysis. This section presents four of these formulas.

3.3.1 FAILURE DENSITY FUNCTION

This is expressed by

$$f(t) = -\frac{dR(t)}{dt} \tag{3.2}$$

where t is time, $f(t)$ is the failure (or probability) density function, and $R(t)$ is the item reliability at time t.

Example 3.1
Assume that the reliability of an item is defined by the following function:

$$R(t) = e^{-\lambda t} \tag{3.3}$$

where λ is the item's constant failure rate.
  Obtain an expression for the item's failure density function.
  Substituting Equation 3.3 into Equation 3.2 yields

$$f(t) = -\frac{d\,e^{-\lambda t}}{dt} \tag{3.4}$$

$$= \lambda\,e^{-\lambda t}$$

3.3.2 HAZARD RATE FUNCTION

This is defined by

$$\lambda(t) = \frac{f(t)}{R(t)} \tag{3.5}$$

where $\lambda(t)$ is the item hazard rate or time-dependent failure rate.
  By inserting Equation 3.2 into Equation 3.5 we get

$$\lambda(t) = -\frac{1}{R(t)} \cdot \frac{dR(t)}{dt} \tag{3.6}$$

Example 3.2

Using Equation 3.3, obtain an expression for the item's hazard rate and comment on the resulting expression.

Substituting Equation 3.3 into Equation 3.6 yields

$$\lambda(t) = -\frac{1}{e^{-\lambda t}} \frac{d\,e^{-\lambda t}}{dt}$$
$$= \lambda \tag{3.7}$$

Thus, the item's hazard rate is given by Equation 3.7.

As the right side of Equation 3.7 is not the function of time t, λ is known as the constant failure rate because it does not depend on time.

3.3.3 GENERAL RELIABILITY FUNCTION

This can be obtained by using Equation 3.6. Thus, rearranging Equation 3.6, we get

$$-\lambda(t)\,dt = \frac{1}{R(t)} \cdot dR(t) \tag{3.8}$$

Integrating both sides of Equation 3.8 over the time interval $[0, t]$, we get

$$-\int_0^t \lambda(t)\,dt = \int_1^{R(t)} \frac{1}{R(t)} \cdot dR(t) \tag{3.9}$$

because at $t = 0$, $R(t) = 1$.

Evaluating the right-hand side of Equation 3.9 yields

$$\ln R(t) = -\int_0^t \lambda(t)\,dt \tag{3.10}$$

Thus, from Equation 3.10, we get the following general expression for reliability function:

$$R(t) = e^{-\int_0^t \lambda(t)\,dt} \tag{3.11}$$

Equation 3.11 can be used to obtain the reliability of an item when its times to failure follow any time-continuous probability distribution.

Example 3.3

Assume that the time to failures of an automobile is exponentially distributed and its failure rate is 0.003 failures per hour. Calculate the automobile's reliability for a 10-hour mission.

Using the data values in Equation 3.11 yields

$$R(10) = e^{-\int_0^{10} (0.003)\, dt}$$
$$= e^{-(0.003)(10)}$$
$$= 0.9704$$

This means there is an approximately 97% chance that the automobile will not fail during the 10-hour mission. More specifically, its reliability will be 0.9704.

3.3.4 Mean Time to Failure

This is an important reliability measure and it can be obtained by using any of the following three formulas [4,5]:

$$MTTF = \int_0^\infty R(t)\, dt \tag{3.12}$$

or

$$MTTF = \int_0^\infty t\, f(t)\, dt \tag{3.13}$$

or

$$MTTF = \lim_{s \to 0} R(s) \tag{3.14}$$

where s is the Laplace transform variable, $MTTF$ is the mean time to failure, and $R(s)$ is the Laplace transform of the reliability function $R(t)$.

Example 3.4

Prove by using Equation 3.3 that Equation 3.12 to Equation 3.14 yield the same result for MTTF.

Thus, by inserting Equation 3.3 into Equation 3.12, we get

$$MTTF = \int_0^\infty e^{-\lambda t}\, dt$$
$$= \frac{1}{\lambda} \tag{3.15}$$

Substituting Equation 3.3 into Equation 3.2 yields

$$f(t) = -\frac{de^{-\lambda t}}{dt}$$
$$= \lambda e^{-\lambda t}$$
(3.16)

Thus, substituting Equation 3.16 into Equation 3.13 yields

$$MTTF = \int_0^\infty t \lambda e^{-\lambda t}\, dt$$
$$= \left[-t\, e^{-\lambda t}\right]_0^\infty - \left[-\frac{e^{-\lambda t}}{\lambda}\right]_0^\infty$$
$$= \frac{1}{\lambda}$$
(3.17)

Taking the Laplace transform of Equation 3.3, we get

$$R(s) = \int_0^\infty e^{-st} \cdot e^{-\lambda t}\, dt$$
$$= \frac{1}{s + \lambda}$$
(3.18)

Substituting Equation 3.18 into Equation 3.14 yields

$$MTTF = \lim_{s \to 0} \frac{1}{(s + \lambda)}$$
$$= \frac{1}{\lambda}$$
(3.19)

Equation 3.15, Equation 3.17, and Equation 3.19 are identical, proving that Equation 3.12 to Equation 3.14 give the same result.

3.4 RELIABILITY NETWORKS

A system can form various configurations in performing reliability analysis. This section is concerned with the reliability evaluation of such commonly occurring configurations or networks.

FIGURE 3.2 A k-unit series system.

3.4.1 SERIES NETWORK

This is probably the most commonly occurring configuration in engineering systems, and its block diagram is shown in Figure 3.2. The diagram represents a k-unit system, and each block in the diagram denotes a unit. All units must work normally for the successful operation of the series system.

The series system (shown in Figure 3.2) reliability is expressed by

$$R_s = P(E_1 \, E_2 \, E_3 \,......E_k)$$ (3.20)

where E_j denotes the successful operation (i.e., success event) of unit j for $j = 1, 2, 3, ..., k$; R_s is the series system reliability; and $P(E_1 \, E_2 \, E_3 \,...... E_k)$ is the occurrence probability of events $E_1, E_2, E_3, ...,$ and E_k.

For independently failing units, Equation 3.20 becomes

$$R_s = P(E_1) P(E_2) P(E_3)......P(E_k)$$ (3.21)

where $P(E_j)$ is the probability of occurrence of event E_j for $j = 1, 2, 3, ..., k$.

If we let $R_j = P(E_j)$ for $j = 1, 2, 3,..., k$ in Equation 3.21 becomes

$$R_s = R_1 \, R_2 \, R_3 \,......R_k$$ (3.22)

where R_j is the unit j reliability for $j = 1, 2, 3, ..., k$.

For the constant failure rate λ_j of unit j from Equation 3.11 (i.e., for $\lambda_j(t) = \lambda_j$), we get

$$R_j(t) = e^{-\lambda_j t}$$ (3.23)

where $R_j(t)$ is the reliability of unit j at time t.

Substituting Equation 3.23 into Equation 3.22 yields

$$R_s(t) = e^{-\sum_{j=1}^{k} \lambda_j t}$$ (3.24)

where $R_s(t)$ is the series system reliability at time t.

Substituting Equation 3.24 into Equation 3.12 yields

$$MTTF_s = \int_0^\infty e^{-\sum_{j=1}^{k} \lambda_j t} \, dt$$

$$= \frac{1}{\sum_{j=1}^{k} \lambda_j}$$

(3.25)

where $MTTF_s$ is the series system mean time to failure.

Example 3.5

Assume that a system is composed of five independent and identical subsystems in series. The constant failure rate of each subsystem is 0.0025 failures per hour. Calculate the reliability of the system for a 50-hour mission and the system mean time to failure.

By substituting the given data into Equation 3.24 we get

$$R_s(100) = e^{-(0.0125)(50)}$$

$$= 0.5353$$

Using the specified data values in Equation 3.25 yields

$$MTTF_s = \frac{1}{5(0.0025)}$$

$$= 80 \; hours$$

Thus, the system reliability and mean time to failure are 0.5353 and 80 hours, respectively.

3.4.2 PARALLEL NETWORK

In this case, the system is composed of k simultaneously operating units, and at least one of these units must operate normally for system success. The block diagram of a k-unit parallel system is shown in Figure 3.3, and each block in the diagram represents a unit.

The parallel system (shown in Figure 3.3) failure probability is given by

$$F_{ps} = P\left(\bar{E}_1 \bar{E}_2 \dots \bar{E}_k\right)$$

(3.26)

where F_{ps} is the parallel system failure probability, $\bar{E}_j$ denotes the failure (i.e., failure event) of unit j, for $j = 1, 2, \dots, k$, and $P\left(\bar{E}_1 \bar{E}_2 \bar{E}_3 \dots \bar{E}_k\right)$ is the occurrence probability of events $\bar{E}_1, \bar{E}_2, \dots, $ and $\bar{E}_k$.

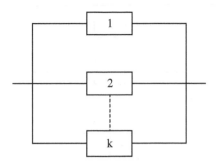

FIGURE 3.3 Block diagram of a *k*-unit parallel system.

For independently failing parallel units, Equation 3.26 becomes

$$F_{ps} = P(\bar{E}_1)\, P(\bar{E}_2)\, ...P(\bar{E}_k) \qquad (3.27)$$

where $P(\bar{E}_j)$ is the probability of occurrence of event $\bar{E}_j$ for $j = 1, 2, ..., k$.
If we let $F_j = P(\bar{E}_j)$ for $j = 1, 2, ..., k$, Equation 3.27 becomes

$$F_{ps} = F_1 F_2 .... F_k \qquad (3.28)$$

where F_j is the unit j failure probability for $j = 1, 2, ..., k$.
By subtracting Equation 3.28 from unity we get

$$\begin{aligned} R_{ps} &= 1 - F_{ps} \\ &= 1 - F_1 F_2 .....F_k \end{aligned} \qquad (3.29)$$

where R_{ps} is the parallel system reliability.
For constant failure rate λ_j of unit j, subtracting Equation 3.23 from unity and then substituting it into Equation 3.29 yields

$$R_{ps}(t) = 1 - \left(1 - e^{-\lambda_1 t}\right)\left(1 - e^{-\lambda_2 t}\right) ...... \left(1 - e^{-\lambda_k t}\right) \qquad (3.30)$$

where $R_{ps}(t)$ is the parallel system reliability at time t.
For identical units, substituting Equation 3.30 into Equation 3.12 yields

$$\begin{aligned} MTTF_{ps} &= \int_0^\infty \left[1 - \left(1 - e^{-\lambda t}\right)^k \right] dt \\ &= \frac{1}{\lambda} \sum_{j=1}^{k} \frac{1}{j} \end{aligned} \qquad (3.31)$$

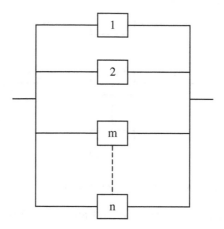

FIGURE 3.4 Block diagram of the m-out-of-n unit system.

where $MTTF_{ps}$ is the parallel system mean time to failure and λ is the unit constant failure rate.

Example 3.6

A system is composed of three independent and identical subsystems. At least one of the subsystems must operate normally for the system to work successfully. Calculate the system's reliability if each subsystem's probability of failure is 0.1.

By substituting the given data into Equation 3.29 we get

$$R_{ps} = 1 - (0.1)(0.1)(0.1)$$
$$= 0.999$$

Thus, the system's reliability is 0.999.

3.4.3 M-Out-of-N Network

In this case, the system is composed of a total of n active units, and least m units must operate normally for system success. The block diagram of an m-out-of-n unit system is shown in Figure 3.4, and each block in the diagram denotes a unit. The series and parallel networks are special cases of the m-out-of-n networks for $m = n$ and $m = 1$, respectively.

For independent and identical units, and using the binomial distribution, we write down the following reliability expression for the Figure 3.4 diagram:

$$R_{m/n} = \sum_{j=m}^{n} \binom{n}{j} R^j (1-R)^{n-j} \tag{3.32}$$

where

$$\binom{n}{j} = \frac{n!}{(n-j)!\,j!} \tag{3.33}$$

where R is the unit reliability and $R_{m/n}$ is the m-out-of-n network reliability.

For constant failure rates of the identical units, substituting Equation 3.3 into Equation 3.32 yields

$$R_{m/n}(t) = \sum_{j=m}^{n} \binom{n}{j} e^{-j\lambda t} \left(1 - e^{-\lambda t}\right)^{n-j} \tag{3.34}$$

where $R_{m/n}(t)$ is the m-out-of-n network reliability at time t and λ is the unit failure rate.

Substituting Equation 3.34 in Equation 3.12 yields

$$MTTF_{m/n} = \int_{0}^{\infty} \left[\sum_{j=m}^{n} \binom{n}{j} e^{-j\lambda t} \left(1 - e^{-\lambda t}\right)^{n-j} \right] dt$$

$$= \frac{1}{\lambda} \sum_{j=m}^{n} \frac{1}{j} \tag{3.35}$$

where $MTTF_{m/n}$ is the m-out-of-n network mean time to failure.

Example 3.7

Assume that an engineering system is composed of four independent and identical units in parallel. At least three units must operate normally for system success. Calculate the system mean time to failure if the unit failure rate is 0.0035 failures per hour.

By substituting the specified data values into Equation 3.35 we get

$$MTTF_{m/n} = \frac{1}{(0.0035)} \sum_{j=3}^{4} \frac{1}{j}$$

$$= \frac{1}{(0.0035)} \left[\frac{1}{3} + \frac{1}{4} \right]$$

$$= 166.67 \ hours$$

Thus, the system mean time to failure is 166.67 hours.

3.4.4 STANDBY SYSTEM

This is another important reliability configuration in which only one unit operates and k units are kept in their standby mode. More specifically, the system contains a total of $k+1$ units, and as soon as the operating unit fails, the switching mechanisms

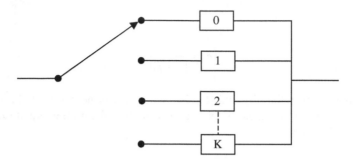

FIGURE 3.5 Block diagram of a standby system with one operating and k standby units.

or other means detect the failure and then replace the failed unit with one of the standby units. Figure 3.5 shows the block diagram of a standby system with one operating and k standby units. Each block in the diagram denotes a unit.

Using the Figure 3.5 diagram for independent and identical units, perfect detection, switching mechanisms and standby units, and time-dependent unit failure rate, we write down the following expression for system reliability [6]:

$$R_{sb}(t) = \sum_{j=0}^{K} \left[\left[\int_{0}^{t} \lambda(t)\, dt \right]^{j} e^{-\int_{0}^{t} \lambda(t)\, dt} \right] / j! \qquad (3.36)$$

where $R_{sb}(t)$ is the standby system reliability at time t and $\lambda(t)$ is the unit time-dependent failure rate.

For constant unit failure rate, (i.e., $\lambda(t) = \lambda$), Equation 3.36 becomes

$$R_{sb}(t) = \sum_{j=0}^{K} (\lambda t)^{j}\, e^{-\lambda t} / j! \qquad (3.37)$$

Inserting Equation 3.37 into Equation 3.12 yields

$$MTTF_{sb} = \int_{0}^{\infty} \left[\sum_{j=0}^{K} (\lambda t)^{j}\, e^{-\lambda t} / j! \right] dt \qquad (3.38)$$

$$= \frac{K+1}{\lambda}$$

where $MTTF_{sb}$ is the standby system mean time to failure.

Example 3.8

A standby system is composed of two independent and identical units: one operating and the other on standby. The unit constant failure rate is 0.0045 failures per hour.

Calculate the system reliability for a 100-hour mission and mean time to failure if the standby unit remains as good as new in its standby mode and failure detection and unit replacement mechanisms are 100% reliable.

By substituting the given data into Equation 3.37 we get

$$R_{sb}(100) = \sum_{j=0}^{K} \left[(0.0045)(100) \right]^j e^{-(0.0045)(100)} / j!$$
$$= 0.9246$$

Using the specified data values in Equation 3.38 yields

$$MTTF_{sb} = \frac{1+1}{(0.0045)} = 444.44 \, hours$$

Thus, the standby system reliability and mean time to failure are 0.9246 and 444.44 hours, respectively.

3.4.5 BRIDGE NETWORK

Sometimes units of an engineering system may form a bridge configuration as shown in Figure 3.6. The diagram is composed of five blocks, each of which denotes a unit. All blocks are labelled with numerals.

For independently failing units, the Figure 3.6 diagram reliability is expressed by

$$R_b = 2 R_1 R_2 R_3 R_4 R_5 + R_2 R_3 R_4 + R_1 R_3 R_5 + R_1 R_4 + R_2 R_5$$
$$- R_2 R_3 R_4 R_5 - R_1 R_2 R_3 R_4 - R_1 R_2 R_3 R_5 - R_3 R_4 R_5 R_1 - R_1 R_2 R_4 R_5$$

(3.39)

where R_i is the reliability of unit i for $i = 1, 2, 3, 4, 5$ and R_b is the bridge network reliability.

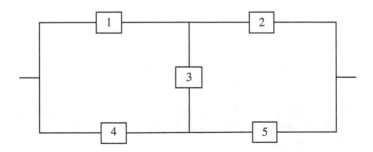

FIGURE 3.6 A bridge network made up of five nonidentical units.

For identical units, Equation 3.39 becomes

$$R_b = 2R^5 - 5R^4 + 2R^3 + 2R^2 \qquad (3.40)$$

For constant unit failure rate, substituting Equation 3.3 into Equation 3.40 yields

$$R_b(t) = 2\,e^{-5\lambda t} - 5e^{-4\lambda t} + 2e^{-3\lambda t} + 2\,e^{-2\lambda t} \qquad (3.41)$$

where $R_b(t)$ is the bridge network reliability at time t and λ is the unit constant failure rate.

By substituting Equation 3.41 into Equation 3.12, we get

$$MTTF_b = \frac{49}{60\lambda} \qquad (3.42)$$

where $MTTF_b$ is the bridge network mean time to failure.

Example 3.9

A system has five independent and identical units forming a bridge configuration. The unit failure rate is 0.0075 failures per hour. Calculate the network reliability for a 100-hour mission and mean time to failure.

Using the given data values in Equation 3.41 yields

$$R_b(100) = 2\,e^{-5\,(0.0075)\,(100)} - 5\,e^{-4\,(0.0075)\,(100)} + 2\,e^{-3\,(0.0075)(100)} + 2\,e^{-2(0.0075)(100)}$$
$$= 0.4552$$

By substituting the specified data value into Equation 3.42, we get

$$MTTF_b = \frac{49}{60(00.0075)}$$
$$= 108.89\ hours$$

Thus, the bridge network's reliability and mean time to failure are 0.4552 and 108.89 hours, respectively.

3.5 RELIABILITY ALLOCATION

This is the process of assigning reliability requirements to individual components for achieving the specified system reliability. Although there are many benefits of the reliability allocation, two of the important ones are as follows [1,7]:

- It forces people involved in design and development to understand and establish the appropriate relationships between reliabilities of components and parts, subsystems, and systems.
- It forces design engineers to consider reliability equally with other design parameters such as cost, performance, and weight.

Over the year, many reliability allocation methods have been developed [8–12]. One of the commonly used methods in the industrial sector is described below [1].

3.5.1 HYBRID METHOD

This method is the result of combining two reliability allocation approaches known as the similar familiar systems method and the factors of influence method. The resulting approach incorporates benefits of these two methods; thus, it is more attractive to use.

The basis for the similar familiar systems method is the designer's familiarity with similar systems or subsystems. More specifically, during the allocation process the method uses the failure data collected on similar systems, subsystems, and items from various sources. The main disadvantage of this approach is to assume that life cycle cost and reliability of past similar designs were satisfactory.

The factors of influence method is based on the assumption that the factors shown in Figure 3.7 effect the system reliability. These are failure criticality, environment, complexity and time, and the state of the art. The failure criticality factor considers the criticality of the failure of the item in question on the system. For example, the failure of some auxiliary instrument in an aircraft may not be as critical as the engine failure.

The environment factor takes into account the exposure or susceptibility of the item or items in question to environmental conditions such as vibration, temperature, and humidity. The complexity and time factor relates to the number of subsystem parts and the relative operating time of the item during the functional period of the complete system.

The state-of-the-art factor takes into account the advancement made in the state-of-the-art for a certain item.

In using the above four factors, every item under consideration is rated with respect to each of these factors by being assigned a number from 1 to 10. The

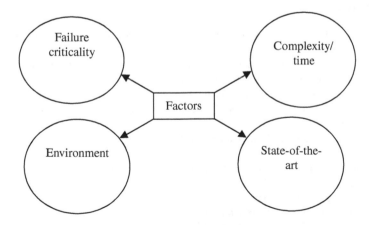

FIGURE 3.7 Factors affecting system reliability.

assignment of 10 means the item under consideration is most affected by the influence factor in question, and 1 means the item is least affected by the same factor.
Subsequently, the reliability is allocated on the basis of the weight of these assigned
numbers for all influence factors considered.

Finally, reliability of an item is allocated by giving certain weights to both
similar familiar systems and factors of influence methods. The hybrid method is
more effective than both these methods used alone because it uses data from both
of them.

3.6 PROBLEMS

1. Write an essay on the need for reliability.
2. Describe the bathtub hazard rate concept.
3. Write down a hazard rate function that can be used to represent a bathtub
   hazard rate curve.
4. Define hazard rate.
5. What is the difference between hazard rate and constant failure rate?
6. Define failure density function.
7. What are the three mathematical approaches for obtaining an item's mean
   time to failure?
8. Discuss reliability allocation and its benefits.
9. Using Equation 3.24, obtain an expression for hazard rate and comment
   on the resulting expression.
10. A system is composed of three independent and identical units in parallel.
    At least two units must operate normally for system success. Calculate
    the system mean time to failure if the unit failure rate is 0.0025 failures
    per hour.

REFERENCES

1. Dhillon, B.S., *Design Reliability: Fundamentals and Applications*, CRC Press, Boca
   Raton, FL, 1999.
2. Kapur, K.C., Reliability and maintainability, in *Handbook of Industrial Engineering*,
   Salvendy, G., Ed., John Wiley & Sons, New York, 1982, pp. 8.5.1–8.5.34.
3. Dhillon, B.S., A hazard rate model, *IEEE Transactions on Reliability*, 28, 150, 1979.
4. Shooman, M.L., *Probabilistic Reliability: An Engineering Approach*, McGraw-Hill,
   New York, 1968.
5. Dhillon, B.S., *Reliability, Quality, and Safety for Engineers*, CRC Press, Boca Raton,
   FL, 2005.
6. Sandler, G.H., *System Reliability Engineering*, Prentice Hall, Englewood Cliffs, NJ,
   1963.
7. Grant Ireson, W., Coombs, C.F., and Moss, R.Y., Eds., *Handbook of Reliability
   Engineering and Management*, McGraw-Hill, New York, 1996.
8. Frederick, H.E., A reliability prediction technique, *Proceedings of the Fourth National
   Symposium on Reliability and Quality Control*, 314–317, 1958.

9. Chipchak, J.S., A practical method of maintainability allocation, *IEEE Transactions on Aerospace and Electronic Systems*, 7, 585–589, 1971.
10. Dhillon, B.S., *Systems Reliability, Maintainability, and Management*, Petrocelli Books, New York, 1983.
11. Balaban, H.S., Jeffers, H.R., and Baechler, D.O., *The Allocation of System Reliability*, Publication No. 152-2-274, ARINC Research Corporation, Chicago, 1961.
12. Von Alven, W.H., Ed., *Reliability Engineering*, Prentice Hall, Englewood Cliffs, NJ, 1964.

4 Reliability Evaluation Tools

4.1 INTRODUCTION

Reliability engineering is wide in scope and it has applications across areas such as aerospace, defense, electric power generation, transportation, and health care. Researchers in these areas have been working to advance the field of reliability. Reliability evaluation is a critical element of reliability engineering and is concerned with ensuring the reliability of engineering products. It usually begins in the conceptual design phase of products with specified reliability.

Over the years, reliability researchers and others working in various areas of reliability engineering have developed many reliability evaluation methods and techniques. Some examples of these methods and techniques are failure modes and effect analysis (FMEA), the network reduction method, the decomposition method, the delta–star method, and the supplementary variables method. The application of these methods and techniques depends on factors such as the type of project under consideration, the specific need, the inclination of the parties involved, and the ease of use.

This chapter presents some of the commonly used reliability evaluation methods and techniques.

4.2 FAILURE MODES AND EFFECT ANALYSIS (FMEA)

This method is a widely used to analyze engineering systems with respect to reliability and it may be simply described as an approach for performing analysis of each system failure mode to examine their effects on the total system [1]. When FMEA is extended to categorize the effect of each potential failure according to its severity, the method is called failure mode effects and criticality analysis (FMECA) [2].

The history of FMEA may be traced back to the early 1950s and the development of flight control systems when the U.S. Navy's Bureau of Aeronautics, in order to establish a mechanism for reliability control over the systems' design effort, developed a requirement known as failure analysis [3]. Subsequently, failure analysis became known as failure effect analysis and then failure modes and effect analysis [4]. The National Aeronautics and Space Administration (NASA) extended FMEA for categorizing the effect of each potential failure according to its severity and called it FMECA [5].

Subsequently, the U.S. Department of Defense developed a military standard entitled "Procedures for Performing a Failure Mode, Effects, and Criticality Analysis." A comprehensive list of publications on FMEA/FMECA is available in Reference 6.

Seven steps are involved in performing FMEA. These are as follows [7]:

- Define system boundaries and associated requirements in detail.
- List all system parts and components and subsystems.
- List all possible failure modes and describe and identify the component or part under consideration.
- Assign appropriate failure rate or probability to each component or part failure mode.
- List effects of each failure mode on subsystems and the plant.
- Enter appropriate remarks for each failure mode.
- Review each critical failure mode and take appropriate action.

This method is described in detail in Reference 8.

4.3 NETWORK REDUCTION METHOD

This is a simple and useful method for determining the reliability of systems consisting of independent series and parallel subsystems. The method sequentially reduces the series and parallel configurations to equivalent units until the whole system becomes a single hypothetical unit [2]. The primary advantage of this method is that it is easy to understand and use. The following example demonstrates the method.

Example 4.1

A network made up of four independents representing a system is shown in Figure 4.1a. Each block in the figure denotes a unit. The reliability of R_i of unit i; for $i = 1, 2, 3, 4$, is given. Calculate the network reliability by using the network reduction method.

First we have identified subsystems A and B of the network shown in Figure 4.1a. Subsystem A is composed of two units in series, and we reduce it to a single hypothetical unit as follows:

$$R_a = R_1 R_2$$

$$R_A = (0.8)(0.4)$$
$$= 0.32$$

where R_A is the reliability of subsystem A.

The reduced network is shown in Figure 4.1b. This network is composed of a parallel subsystem (i.e., subsystem B) in series with a single unit. Thus, we reduce the parallel subsystem to a single hypothetical unit as follows:

$$R_B = 1 - (1 - R_A)(1 - R_3)$$
$$= 1 - (1 - 0.32)(1 - 0.7)$$
$$= 0.796$$

where R_B is the reliability of subsystem B.

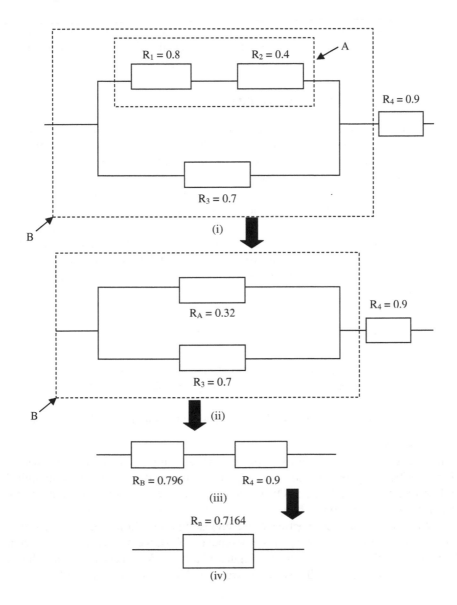

FIGURE 4.1 Diagrammatic steps of the network reduction method: (i) original network, (ii) reduced network, (iii) further reduced network, and (iv) single hypothetical unit.

The reduced network with the above calculated value is shown in Figure 4.1c. This is a two-unit series network, and its reliability is given by

$$R_n = R_B R_4$$
$$= (0.796)(0.9)$$
$$= 0.7164$$

where R_n is the whole-network reliability.

The single hypothetical unit shown in Figure 4.1d represents the reliability of the whole network given in Figure 4.1a. More specifically, the entire network is reduced to a single hypothetical unit, and its reliability is 0.7164.

4.4 DECOMPOSITION METHOD

This is a quite useful method used to evaluate reliability of complex systems. It decomposes a complex system into simpler subsystems by applying the conditional probability theory. The system reliability is obtained by combining the reliability measures of subsystems [9]. The basis for the method is the selection of the key element or unit used for decomposing a given network representing a complex system. The efficiency of the technique depends on the selection of this key element.

This approach begins by assuming that the key element, for example, z, is replaced by another element that never fails (i.e., 100% reliable) and it assumes that the key element is completely removed from the complex system under consideration. The overall reliability of the complex system is obtained by using the following equation [9]:

$$R_{CS} = P(z)\,P(system\ good\ /\ z\ good) + P\left(\overline{z}\right)P(system\ good\ /\ z\ fails) \qquad (4.1)$$

where R_{CS} is the complex system or network reliability, $P(\cdot)$ is the probability, $P(z)$ is the reliability of the key element z, and $P\left(\overline{z}\right)$ is the failure probability of the key element z.

The application of this approach is demonstrated through the following example.

Example 4.2

An independent unit network representing a complex system is shown in Figure 4.2a. Each block and letter R_i in the figure denote a unit and unit i reliability, for $i = 1, 2, \ldots, 5$. Obtain an expression for the network reliability by using the decomposition technique.

With the aid of past experience, we choose the unit falling between nodes C and D shown in Figure 4.2a as our key element z. Next, we replace the key element with a bad (failed) element; thus, the Figure 4.2a network reduces to the one shown in Figure 4.2b. It is a parallel-series network, and its reliability is given by

$$R_{ps} = 1 - (1 - R_1\,R_4)(1 - R_2\,R_5) \qquad (4.2)$$

where R_{ps} is the parallel-series network reliability.

Similarly, we replace the key element with a perfect element that never fails; thus, the Figure 4.2a network reduces to the one shown in Figure 4.2c. It is a series-parallel network, and its reliability is given by

$$R_{sp} = \left[1 - \left(1 - R_1\right)\left(1 - R_2\right)\right]\left[1 - \left(1 - R_4\right)\left(1 - R_5\right)\right] \qquad (4.3)$$

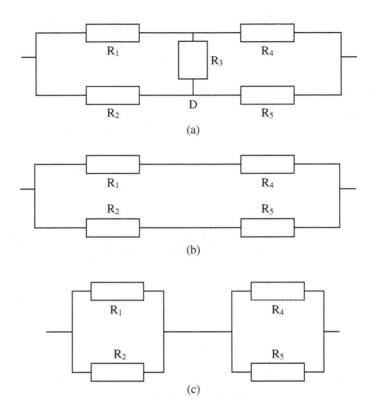

FIGURE 4.2 (a) complex network; (b) reduced network because of the bad (failed) key element, (c) reduced network because of the perfect (i.e., 100% reliable) key element.

where R_{sp} is the series-parallel network reliability.

The reliability and failure probability, respectively, of the key element z are given by

$$P(z)=R_3 \qquad (4.4)$$

and

$$P(\overline{z})=1-R_3 \qquad (4.5)$$

where R_3 is the reliability of the key element z.

By substituting Equation 4.2 to Equation 4.5 into Equation 4.1, we get

$$R_N = R_3 R_{sp} + (1-R_3) R_{ps}$$
$$= R_3 \left[1-(1-R_1)(1-R_2)\right]\left[1-(1-R_4)(1-R_5)\right]+(1-R_3)$$
$$\left[1-(1-R_1 R_4)(1-R_2 R_5)\right] \qquad (4.6)$$

where R_N is the reliability of the network shown in Figure 4.2a.

For identical units (i.e., $R_1 = R_2 = R_3 = R_4 = R_5 = R$), Equation 4.6 simplifies to

$$R_N = 2R^5 - 5R^4 + 2R^3 + 2R^2 \qquad (4.7)$$

4.5 DELTA–STAR METHOD

This is a very practical and powerful method to determine the reliability of bridge networks [10]. This approach transforms a bridge network to its equivalent network composed of series and parallel configurations. However, the transformation process introduces a small error in the end result, but for practical applications this error should be considered negligible [2].

After the transformation of a bridge network to its equivalent series and parallel form, the network reduction method presented earlier in the chapter can be used to obtain network reliability. Nonetheless, the delta–star approach can easily handle complex networks containing more than one bridge configuration as well as bridge networks composed of units or devices with two mutually exclusive failure modes [2].

Figure 4.3 shows a delta–star equivalent reliability diagram. Each block in the figure denotes the respective unit reliability ($R_{(\cdot)}$) and the nodes (numerals). The reliabilities of units falling between nodes 1 and 2, 3 and 2, and 3 and 1 in the delta configuration in Figure 4.3 are R_{12}, R_{32}, and R_{31}, respectively.

Similarly, the reliabilities of units close to nodes 1, 2, and 3 in the star configuration in Figure 4.3 are R_1, R_2, and R_3, respectively.

Using Figure 4.3, we write down the following equivalent reliability equations for independent unit networks falling between nodes 1 and 2, 3 and 2, and 3 and 1, respectively:

$$R_1 R_2 = 1 - \left(1 - R_{31} R_{32}\right)\left(1 - R_{12}\right) \qquad (4.8)$$

$$R_3 R_2 = 1 - \left(1 - R_{31} R_{12}\right)\left(1 - R_{32}\right) \qquad (4.9)$$

$$R_3 R_1 = 1 - \left(1 - R_{12} R_{32}\right)\left(1 - R_{31}\right) \qquad (4.10)$$

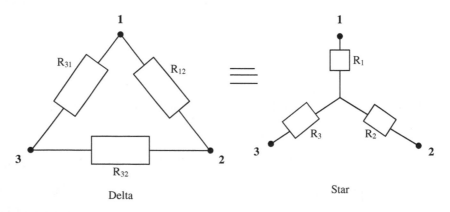

FIGURE 4.3 A delta–star equivalent reliability diagram.

Solving Equation 4.8 to 4.10, we get

$$R_1 = \left[\frac{AC}{B} \right]^{1/2} \tag{4.11}$$

$$R_2 = \left[\frac{AB}{C} \right]^{1/2} \tag{4.12}$$

$$R_3 = \left[\frac{BC}{A} \right]^{1/2} \tag{4.13}$$

where

$$A = 1 - \left(1 - R_{31} R_{32}\right)\left(1 - R_{12}\right) \tag{4.14}$$

$$B = 1 - \left(1 - R_{31} R_{12}\right)\left(1 - R_{32}\right) \tag{4.15}$$

$$C = 1 - \left(1 - R_{12} R_{32}\right)\left(1 - R_{31}\right) \tag{4.16}$$

Example 4.3

A bridge network made up of five independent and identical units is shown in Figure 4.4. Each block in the figure denotes a unit with specified reliability $R = 0.9$. Calculate the network reliability by using the delta–star method. Use the same unit specified reliability value (i.e., $R = 0.9$) in Equation 4.7 to obtain the bridge network reliability. Comment on both results.

In Figure 4.4 nodes labeled 1, 2, and 3 denote the delta configuration. Using the given data in Equation 4.11 to Equation 4.13, we get the following star equivalent

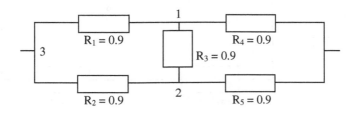

FIGURE 4.4 A five identical-unit bridge network.

reliabilities for the delta configuration:

$$R_1 = \left[\frac{AC}{B}\right]^{1/2}$$
$$= 0.9904$$

where

$$A = B = C = 1 - \left[1 - (0.9)(0.9)\right](1 - 0.9)$$
$$= 0.981$$
$$R_2 = 0.9904$$
$$R_3 = 0.9904$$

Using the above calculated values, the equivalent network to the Figure 4.4 bridge network is redrawn in Figure 4.5.

The reliability of the Figure 4.5 network, R_n, is given by

$$R_n = R_3\left[1 - (1 - R_1 R_4)(1 - R_2 R_5)\right]$$
$$= (0.9904)\left[1 - \{1 - (0.9904)(0.9)\}\{1 - (0.9904)(0.9)\}\right]$$
$$= 0.9788$$

Using the data from Equation 4.7 yields

$$R_N = 2(0.9)^5 - 5(0.9)^4 + 2(0.9)^3 + 2(0.9)^2$$
$$= 0.9785$$

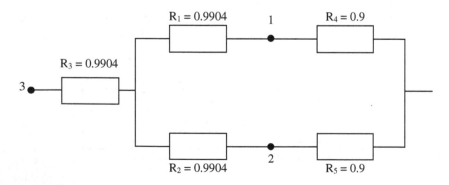

FIGURE 4.5 Equivalent network to the bridge configuration shown in Figure 4.4.

The two above reliability values, obtained by using the exact method and the delta–star, approach are basically same (i.e., 0.9785 and 0.9788). This demonstrates that for practical purposes the delta–star method is quite effective.

4.6 MARKOV METHOD

This is one of the most widely used reliability analysis methods and is named after a Russian mathematician Andrei Andreyevich Markov (1856–1922). It is mainly used to analyze repairable and nonrepairable systems with constant failure/repair rates. The method is subject to the following assumptions [9]:

- The transitional probability from one state to another in the finite time interval Δt is given by $\lambda \Delta t$, where λ is the constant transition rate (i.e., the failure or repair rate) from one system state to another.
- All occurrences are independent of each other.
- The probability of occurrence of more than one transition in finite time interval Δt from one state to another is very small or negligible (e.g., $[\lambda \Delta t]$ $[\lambda \Delta t] \rightarrow 0$).

The following example demonstrates the application of this method [2].

Example 4.4
A mechanical system can be in one of two states: operating normally or failed. Its constant failure and repair rates are λ_m and μ_m, respectively. The system state space diagram is shown in Figure 4.6. The numerals in circles donate the system states.

Obtain expressions, by using the Markov method, for system time-dependent and steady-state availabilities and unavailabilities.

Using the Markov method, we write down the following equations for the Figure 4.6 diagram:

$$P_0(t+\Delta t)=P_0(t)(1-\lambda_m\,\Delta t)+\mu_m\,\Delta t\,P_1(t) \tag{4.17}$$

$$P_1(t+\Delta t)=P_1(t)(1-\mu_m\,\Delta t)+\lambda_m\,\Delta t\,P_0(t) \tag{4.18}$$

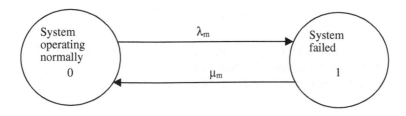

FIGURE 4.6 System state space diagram.

where $P_0(t+\Delta t)$ is the probability that the system is in operating state 0 at time $(t+\Delta t)$, $P_1(t+\Delta t)$ is the probability that the system is in operating state 1 at time $(t+\Delta t)$, t is time, $P_0(t)$ is the probability that the system is in operating state 0 at time t, $P_1(t)$ is the probability that the system is in failed state 1 at time t, $\lambda_m \Delta t$ is the probability of system failure in finite time interval Δt, $\mu_m \Delta t$ is the probability of system repair in infinite time interval Δt, $(1-\lambda_m \Delta t)$ is the probability of no failure in finite time interval Δt, and $(1-\mu_m \Delta t)$ is the probability of no repair in finite time interval Δt.

In the limiting case, Equation 4.17 and Equation 4.18 become

$$\frac{dP_0(t)}{dt} + \lambda_m P_0(t) = \mu_m P_1(t) \tag{4.19}$$

$$\frac{dP_1(t)}{dt} + \mu_m P_1(t) = P_0(t)\lambda_m \tag{4.20}$$

at time $t=0$, $P_0(0)=1$, and $P_1(0)=0$.

By solving Equation 4.19 and Equation 4.20, we get

$$P_0(t) = \frac{\mu_m}{\left(\lambda_m + \mu_m\right)} + \frac{\lambda_m}{\left(\lambda_m + \mu_m\right)} e^{-\left(\lambda_m + \mu_m\right)t} \tag{4.21}$$

$$P_1(t) = \frac{\lambda_m}{\left(\lambda_m + \mu_m\right)} - \frac{\lambda_m}{\left(\lambda_m + \mu_m\right)} e^{-\left(\lambda_m + \mu_m\right)t} \tag{4.22}$$

Thus, the mechanical system time-dependent availability and unavailability, respectively, are

$$A_m(t) = P_0(t) = \frac{\mu_m}{\left(\lambda_m + \mu_m\right)} + \frac{\lambda_m}{\left(\lambda_m + \mu_m\right)} e^{-\left(\lambda_m + \mu_m\right)t} \tag{4.23}$$

and

$$UA_m(t) = P_1(t) = \frac{\lambda_m}{\left(\lambda_m + \mu_m\right)} - \frac{\lambda_m}{\left(\lambda_m + \mu_m\right)} e^{-\left(\lambda_m + \mu_m\right)t} \tag{4.24}$$

where $A_m(t)$ is the availability of the mechanical system at time t and $UA_m(t)$ is the unavailability of the mechanical system at time t.

For large t, Equation 4.23 and Equation 4.24 become

$$A_m = \frac{\mu_m}{\lambda_m + \mu_m} \tag{4.25}$$

and

$$UA_m = \frac{\lambda_m}{\lambda_m + \mu_m} \qquad (4.26)$$

where A_m is the steady-state availability of the mechanical system and UA_m is the steady-state unavailability of the mechanical system.

Since $\lambda_m = 1/MTFF_m$ and $\mu_m = 1/MTRR_m$, Equation 4.25 and Equation 4.26 become

$$A = \frac{MTTF_m}{MTTF_m + MTTR_m} \qquad (4.27)$$

where $MTTF_m$ is the mechanical system mean time to failure and $MTTR_m$ is the mechanical system mean time to repair and

$$UA_m = \frac{MTTR_m}{MTTF_m + MTTR_m} \qquad (4.28)$$

4.7 SUPPLEMENTARY VARIABLES METHOD

This method is used to model systems with constant failure rates and nonexponential repair times [11–13]. These systems are sometimes called the non-Markovian systems since the stochastic process is non-Markovian. The inclusion of sufficient supplementary variables in the specification of the state of the system can make a process Markovian [11]. The application of this method is demonstrated through the following example [13,14].

Example 4.5

An operating system can fail either fully or partially, and from the partially operating state it fails completely. More specifically, the system can be in either of three states: operational, partially operational, or failed. The system is repaired from partially and fully failed states to the normal operating state. The system state space diagram is shown in Figure 4.7 [14]. The numerals in the diagram donate system states.

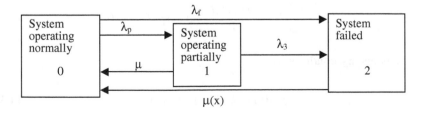

FIGURE 4.7 System state space diagram.

The following assumptions are associated with the system:

- All system failure rates are constant.
- The fully failed system repair times are arbitrarily distributed.
- The partially failed system repair rate is constant.
- All system failures are statistically independent.
- The partially or fully failed system is restored to as good as new condition.

Using the supplementary variables method, obtain Laplace transforms of system state probabilities.

The following symbols associated with the Figure 4.7 diagram are used to develop equations for the model representing the system:

- i denotes the ith state of the system: $i = 0$ (system operating normally), $i = 1$ (system operating partially), $i = 2$ (system failed).
- μ is the system constant repair rate from partial to normal operating state.
- λ_j is the system jth constant failure rate: $j = f$ (normal to fully failed state), $j = p$ (normal to partially operating state), and $j = 3$ (partially operating to fully failed state).
- $P_i(t)$ is the probability that the system is in state i at time t; for $i = 0, 1, 2$.
- $p_2(x, t)\Delta x$ is the probability that at time t, the system is in state 2 and the elapsed repair time lies in the interval $[x, x + \Delta x]$.
- $\mu(x)$ and $q(x)$ are the repair rate and probability density of repair times, respectively, when system is in state 2 and has an elapsed repair time of x.
- s is the Laplace transform variable.

Using Figure 4.7 and the supplementary variables method, we write down the following equations:

$$\frac{dP_0(t)}{dt} + \left(\lambda_p + \lambda_f\right)P_0(t) - P_1(t)\mu = \int_0^\infty p_2(x,t)\mu(x)\,dx \qquad (4.29)$$

$$\frac{dP_1(t)}{dt} + \left(\lambda_3 + \mu\right)P_1(t) - P_0(t)\lambda_p = 0 \qquad (4.30)$$

$$\frac{\partial p_2(x,t)}{\partial x} + \frac{\partial p_2(x,t)}{\partial t} + \mu(x)p_2(x,t) = 0 \qquad (4.31)$$

The boundary condition is

$$p_2(0,t) = \lambda_3 P_1(t) + \lambda_f P_0(t) \qquad (4.32)$$

at $t = 0$, $P_0(0) = 1$, $P_1(0) = 0$, and $p_2(x, 0) = 0$.

By solving Equation 4.29 to Equation 4.32, we get

$$P_0(s)=\left[s+\left(\lambda_p+\lambda_f\right)-\frac{\lambda_p\mu}{(s+\lambda_3+\mu)}-\left(\frac{\lambda_3\lambda_p}{s+\lambda_3+\mu}+\lambda_f\right)G(s)\right]^{-1} \quad (4.33)$$

where
$$G(s)\equiv\int_0^\infty e^{-sx}q(x)dx \quad\text{and}\quad q(x)=\mu(x)e^{-\int_0^x\mu(x)\,dx}.$$

$$P_1(s)\equiv\frac{P_0(s)\lambda_p}{\left(s+\lambda_3+\mu\right)} \quad (4.34)$$

$$P_2(s)=\left[\frac{\lambda_p\lambda_3}{(s+\lambda_3+\mu)}+\lambda_f\right]\left(\frac{1-G(s)}{s}\right)P_0(s) \quad (4.35)$$

Equation 4.33 to Equation 4.35 are the Laplace transforms of the system state probabilities. For a given probability density function, $q(x)$, of failed system repair times, Equation 4.33 to Equation 4.35 can be inverted to obtain expressions for state probabilities $P_0(t)$, $P_1(t)$, and $P_2(t)$.

The supplementary variables method is described in detail in Reference 14.

4.8 PROBLEMS

1. Write an essay on commonly used reliability evaluation methods.
2. Discuss the history of failure modes and effect analysis (FMEA).
3. Discuss differences between FMEA and failure mode effects and critical analysis (FMECA).
4. A network made up of seven independent units representing a system is shown in Figure 4.8. Each unit's reliability, R_i for i = 1, 2, 3, 4, and 5, is given. Calculate the network reliability by using the network reduction method.

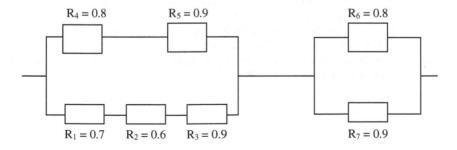

FIGURE 4.8 A seven independent-unit reliability network.

5. What is the decomposition method?
6. What is the main disadvantage of the delta–star method?
7. What are the assumptions associated with the Markov method?
8. What are the differences between the Markov and the supplementary variables methods?
9. What are the advantages of the delta–star method?
10. Assume that the failed system times to repair are exponentially distributed in Equation 4.33 to Equation 4.35. Obtain expressions for the system state probabilities $P_0(t)$, $P_1(t)$, and $P_2(t)$.

REFERENCES

1. Omdahl, T.P., Ed., *Reliability, Availability, and Maintainability (RAM) Dictionary*, American Society for Quality Control (ASQC) Press, Milwaukee, WI, 1988.
2. Dhillon, B.S., *Design Reliability: Fundamentals and Applications*, CRC Press, Boca Raton, FL, 1999.
3. *General Specification for Design, Installation, and Test of Aircraft Flight Control Systems*, MIL-F-18372 (Aer), Bureau of Naval Weapons, Department of the Navy, Washington, DC, Para. 3.5.2.3.
4. Continho, J.S., Failure effect analysis, *Transactions of the New York Academy of Sciences*, Ser II, 26, 564–584, 1963–1964.
5. Jordan, W.E., Failure modes, effects, and criticality analyses, *Proceedings of the Annual Reliability and Maintainability Symposium*, 30–37, 1972.
6. Dhillon, B.S., Failure modes and effects analysis: bibliography, *Microelectronics and Reliability*, 32, 719–731, 1992.
7. Dhillon, B.S. and Singh, C., *Engineering Reliability: New Techniques and Applications*, John Wiley & Sons, New York, 1981.
8. Palady, P., *Failure Modes and Effect Analysis*, PT Publications, West Palm Beach, FL, 1995.
9. Shooman, M.L., *Probabilistic Reliability: An Engineering Approach*, McGraw-Hill, New York, 1968.
10. Dhillon, B.S., The Analysis of the Reliability of Multi-state Device Networks, Ph.D. dissertation, 1975. Available from the National Library of Canada, Ottawa.
11. Cox, D.R., The analysis of non-Markovian stochastic processes, by the inclusion of supplementary variables, *Proceedings of the Cambridge Philosophical Society*, 51, 433–441, 1955.
12. Gaver, D.P., Time to failure and availability of paralleled systems with repair, *IEEE Transactions on Reliability*, 12, 30–38, 1963.
13. Garg, R.C., Dependability of a complex system having two types of components, *IEEE Transactions on Reliability*, 12, 11–15, 1963.
14. Dhillon, B.S., *Reliability Engineering in Systems Design and Operation*, Van Nostrand Reinhold, New York, 1983.

5 Reliability Management

5.1 INTRODUCTION

Reliability management has become an important element of reliability engineering because of various factors including system complexity, sophistication, and size; demanding reliability requirements; and cost and time constraints. Reliability management is concerned with the direction and control of an organization's reliability activities such as developing reliability policies and goals, facilitating interactions of reliability manpower with other parts of the organization, and staffing.

The history of reliability management can be traced back to the late 1950s when the U.S. Air Force developed a reliability program management document (i.e., Exhibit 58-10) [1]. Subsequently, the efforts of the U.S. Department of Defense to develop requirements for an organized contractor reliability program resulted in the release of the military specification MIL-R-27542 [2].

Many publications directly or indirectly relating to reliability management have appeared [3]. This chapter presents various important aspects of reliability management.

5.2 GENERAL MANAGEMENT RELIABILITY PROGRAM RESPONSIBILITIES

General management plays an important role in the success of a reliability program by fulfilling its responsibilities in an effective manner. Some of its responsibilities are as follows [4]:

- Developing appropriate reliability goals.
- Providing appropriate funds, manpower, scheduled time, and authority.
- Establishing an effective program to fulfill set reliability objectives or goals and eradicating existing shortcomings.
- Developing a mechanism for accessing information concerning current reliability performance of the organization with respect to its operations and products.
- Monitoring the program regularly and taking appropriate corrective measures with respect to associated policies, procedures, organization, and so on.

5.3 A METHOD FOR ESTABLISHING RELIABILITY GOALS AND GUIDELINES FOR DEVELOPING RELIABILITY PROGRAMS

When working from preestablished reliability requirements, it is essential to develop appropriate reliability goals. This involves reducing the requirements to a series of subgoals. The steps of a useful method for establishing reliability goals are shown in Figure 5.1 [5].

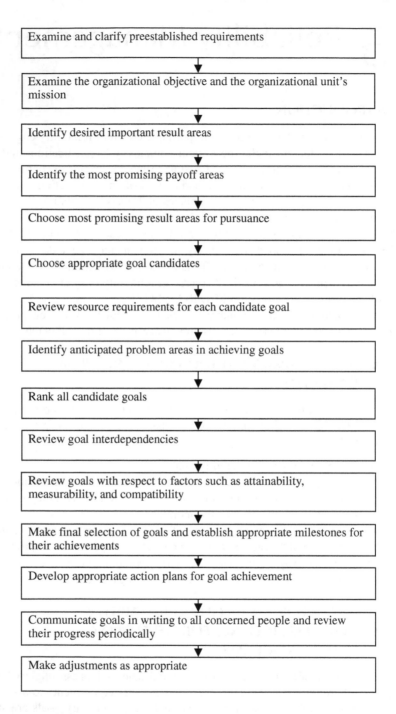

FIGURE 5.1 Steps of a method for developing reliability goals.

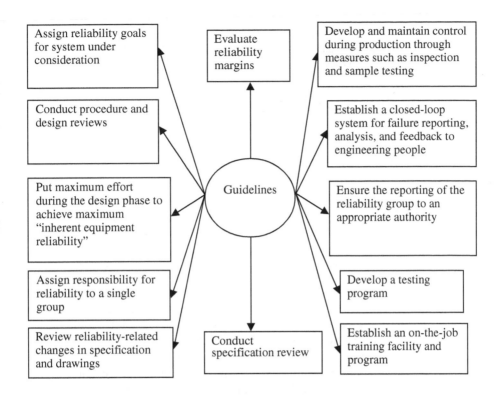

Assign reliability goals for system under consideration

Evaluate reliability margins

Develop and maintain control during production through measures such as inspection and sample testing

Conduct procedure and design reviews

Establish a closed-loop system for failure reporting, analysis, and feedback to engineering people

Put maximum effort during the design phase to achieve maximum "inherent equipment reliability"

Guidelines

Ensure the reporting of the reliability group to an appropriate authority

Assign responsibility for reliability to a single group

Develop a testing program

Review reliability-related changes in specification and drawings

Conduct specification review

Establish an on-the-job training facility and program

FIGURE 5.2 Useful guidelines for developing reliability programs.

Various guidelines have appeared in the published literature for developing reliability programs. Figure 5.2 presents 12 guidelines for developing reliability programs that appeared in the military specification MIL-R-27542 [6].

5.4 RELIABILITY AND MAINTAINABILITY MANAGEMENT TASKS IN SYSTEM LIFE CYCLE

To obtain the desired level of reliability of a system in the field environment, a series of management tasks with respect to reliability and maintainability must be performed throughout the system life cycle. The system life cycle may be divided into four phases as shown in Figure 5.3 [7]. The reliability- and maintainability-related management tasks involved in each of these phases are presented below.

5.4.1 CONCEPT AND DEFINITION PHASE

This is the first phase of the system life cycle, in which system requirements are established and the basic characteristics are defined. During this phase various reliability- and maintainability-related management tasks are performed.

Some of these tasks are: (a) defining the system capability requirements, management controls, parts control requirements, and terms used; (b) defining the

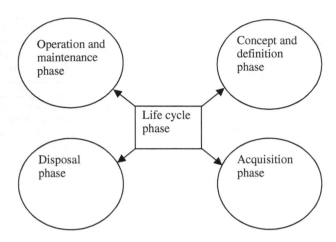

FIGURE 5.3 System life cycle phases.

system reliability and maintainability goals in quantitative terms; (c) defining system environmental factors during its life cycle; (d) defining hardware and software standard documents to be used to meet reliability and maintainability requirements; (e) defining constraints proven to be harmful to reliability; (f) defining data collection and analysis needs during the system life cycle; (g) defining methods to be used during the design and manufacturing phase; (h) defining the management controls required for documentation; (i) defining the combination of machines, facilities, manpower, and tools required to produce the design and its assemblies to the stated specifications; (j) defining system safety requirements; and (k) defining the basic maintenance philosophy.

5.4.2 ACQUISITION PHASE

This phase is concerned with activities related to system acquisition and installation as well as planning for the eventual support of the system under consideration. Many reliability- and maintainability-related management tasks are involved in this phase. Some of these tasks are as follows:

- Define all system technical requirements.
- Define the major design and development methods to be employed.
- Define the demonstration requirements.
- Define all the reliability and maintainability requirements to be satisfied.
- Define all the documents required as part of the final product, system, or equipment.
- Define the kind of evaluation methods to be used to assess the system.
- Define the kind of reviews to be performed.
- Define the kind of data the manufacturer must provide to the customer.
- Define the meaning of a degradation or failure.
- Define the cost restraints and the life cycle cost information to be developed.

TABLE 5.1
Reliability- and Maintainability-Related Management Tasks
Associated with the Operation and Maintenance Phase

No.	Task
1	• Collect, monitor, and analyze reliability and maintainability data
2	• Manage and predict spare parts
3	• Establish failure data banks
4	• Provide adequate maintenance tools and test equipment
5	• Prepare engineering and maintenance documents
6	• Review the documentation with respect to any engineering change
7	• Develop engineering change proposals
8	• Provide adequate manpower
9	• Develop maintenance support for the various levels of maintenance

- Define the type of field studies, if any, to be conducted.
- Define the controls to be exercised by both the manufacturer and the customer during this phase.
- Define the kind of future logistics required (i.e., during initial acquisition and in-service period).

5.4.3 OPERATION AND MAINTENANCE PHASE

This phase is concerned with tasks associated with the maintenance activity, management of the engineering, and support of the system during its entire operational life. Some of the reliability- and maintainability-related management tasks associated with this phase are presented in Table 5.1.

5.4.4 DISPOSAL PHASE

This phase is concerned with tasks that are required to remove the system and all its nonessential supporting material. Two of the reliability- and maintainability-related management tasks involved in this phase are calculating the final system life cycle cost and the reliability and maintainability values. The resulting life cycle cost takes into consideration the disposal action cost or income. The reliability and maintainability values are computed for the buyer of the used system as well as for application in the procurement of similar systems.

5.5 RELIABILITY MANAGEMENT TOOLS AND DOCUMENTS

Reliability management uses a variety of tools and documents. Some examples of these tools are configuration management, value engineering, and critical path method. Similarly, some of the documents used by reliability management are the reliability manual, international and national specifications and standards, policy and procedure documents, plans and instructions, and reports and drawings [7].

Some of the above items are discussed below.

5.5.1 CONFIGURATION MANAGEMENT

During the development of an engineering system or product many changes may occur; these changes may be concerned with product performance, weight, size, appearance, and so on. Configuration management is a useful tool to assure the customer and the manufacturer that the end product will fully satisfy the contract specification. Thus, configuration management is the management of engineering requirements that defines the engineering product or system as well as changes thereto [8].

The history of configuration management can be traced back to 1962, when the U.S. Air Force released a document entitled "Configuration Management During the Development and Acquisition Phases," AFSCM 375-1 [9]. Today configuration management is well known in the industrial sector, and its advantages include reduction in overall cost, effective channeling of resources, facilitation of accurate data retrieval, elimination of redundant efforts, formal establishment of objectives, and precisely identified final product.

Configuration management is described in detail in Reference 10.

5.5.2 VALUE ENGINEERING

Value engineering is a systematic, creative technique used to accomplish a necessary function at the minimum cost [11]. Historical records clearly indicate that the application of the value engineering approach has returned somewhere between $15 and $30 for each dollar spent [12]. The financial returns from the application of this concept are very promising.

The history of the value engineering concept can be traced back to 1947, when General Electric Company assigned Lawrence D. Miles a project to develop methods for reducing costs through material substitution or changes in design or production methods [7].

There are many areas in which value engineering is useful: identifying areas requiring attention and improvement, prioritizing, serving as a vehicle for dialogue, increasing the value of good and services, and generating new ideas to solve problems. It also serves as a useful tool to determine alternative solutions to a concern, a useful procedure for assigning dollars on high-value items, a means to document rationales behind decisions, and a useful approach for determining and quantifying intangibles [12].

Additional information on value engineering is available in Reference 13.

5.5.3 CRITICAL PATH METHOD

The critical path method (CPM) along with the program evaluation and review technique (PERT) is widely used for planning and controlling projects. It was developed by E.I. DuPont de Nemours and Company in 1956 for scheduling

TABLE 5.2
Topics Covered in a Reliability Manual

No.	Topic
1	• Company-wide reliability policy
2	• Organizational structure and responsibilities
3	• Relationship with suppliers and customers
4	• Product design phase procedures from the standpoint of reliability
5	• Effective reliability methods, models, etc.
6	• Reliability test and demonstration approaches and procedures
7	• Failure data collection and analysis methods and procedures to be followed

design- and construction-related activities [14]. The following general steps are associated with this method:

- Break down the project under consideration into individual tasks or jobs.
- Arrange the jobs or tasks into a logical network.
- Estimate the duration time of each task or job.
- Develop a schedule.
- Highlight jobs or tasks that control the completion of the project.
- Redistribute resources and funds to improve the schedule.

Some of the advantages of the CPM are that it is useful to determine project duration systematically, show interrelationships in work flow, improve communication and understanding, identify critical work activities for completing the project on time, monitor project progress effectively, and determine the need for labor and resources in advance. It is also useful in cost control and cost saving [15].

CPM is described in detail in Reference 16.

5.5.4 RELIABILITY MANUAL

This is the backbone of any reliability organization. Its existence is absolutely necessary for any organization irrespective of its size. A typical reliability manual covers topics such as those presented in Table 5.2 [7].

5.6 RELIABILITY DEPARTMENT FUNCTIONS AND TASKS OF RELIABILITY ENGINEER

A reliability department performs a variety of reliability-related functions. Some of these are developing reliability policy, plans, and procedures; providing reliability-related inputs to design proposals and specifications; carrying out reliability allocation and prediction; training reliability manpower; conducting reliability-related research; auditing reliability-related activities; carrying out reliability demonstration and failure data collection and reporting; monitoring the reliability activities of subcontractors (if any), monitoring reliability growth, conducting specification and

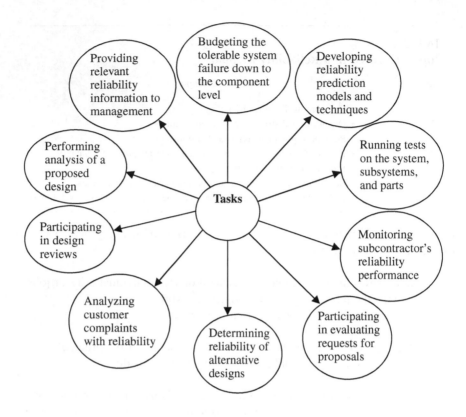

FIGURE 5.4 Important tasks of a reliability engineer.

design reviews with respect to reliability; consulting on reliability matters; and conducting failure data analysis [2].

A reliability engineer performs various types of tasks. Some of the important ones are shown in Figure 5.4 [2].

5.7 PITFALLS IN RELIABILITY PROGRAM MANAGEMENT

Past experiences indicate that many reliability program uncertainties and problems are the result of pitfalls in reliability program management. These pitfalls occur in many areas, as shown in Figure 5.5 [17]. A typical example of the pitfalls that occur in reliability testing is the delay in starting the reliability-demonstrating testing. Similarly, two typical examples of the pitfalls in the reliability organization areas are having several organizational tiers and many persons with authority to make commitments without any dialogue and coordination.

Many programming-related pitfalls require careful attention. One example of these pitfalls is the assumption that each person associated with the program clearly understands the specified reliability requirements.

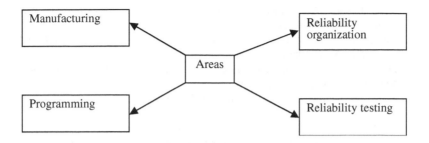

FIGURE 5.5 Main areas of pitfalls in reliability program management.

Finally, an example of the manufacturing phase pitfalls is the authorization of substitute parts, without paying much attention to their effect on reliability, by parts buyers or others when the parts acquisition lead time is incompatible with the system manufacturing schedule.

All in all, the occurrence of pitfalls such as these can only be avoided by an effective reliability management team.

5.8 PROBLEMS

1. Write an essay on reliability management.
2. Discuss the general management reliability program responsibilities.
3. Describe a method for developing reliability goals.
4. List at least ten useful guidelines for developing reliability programs.
5. Discuss reliability and maintainability management tasks for the system acquisition phase.
6. Discuss the following items:
   • Configuration management
   • Value engineering
7. Discuss the topics covered in a reliability manual.
8. List at least ten important functions of a reliability department.
9. Discuss important tasks of a reliability engineer.
10. Discuss reliability program management pitfalls.

REFERENCES

1. Austin-Davis, W., Reliability management: a challenge, *IEEE Transactions on Reliability*, 12, 6–9, 1963.
2. Dhillon, B.S., Engineering reliability management, *IEEE Journal on Selected Areas in Communications*, 4(7), 1015–1020, 1986.
3. Dhillon, B.S., *Reliability and Quality Control: Bibliography on General and Specialized Areas*, Beta Publishers, Gloucester, Ontario, Canada, 1992.
4. Heyel, C., *The Encyclopaedia of Management*, Van Nostrand Reinhold, New York, 1979.

 5. Grant Ireson, W., Coombs, C.F., and Moss, R.Y., Eds., *Handbook of Reliability Engineering and Management*, McGraw-Hill, New York, 1996.
 6. Karger, D.W. and Murdick, R.G., *Managing Engineering and Research*, Industrial Press, New York, 1980.
 7. Dhillon, B.S. and Reiche, H., *Reliability and Maintainability Management*, Van Nostrand Reinhold, New York, 1985.
 8. Feller, M., Configuration management, *IEEE Transactions on Engineering Management*, 16, 64–66, 1969.
 9. *Configuration Management During the Development and Acquisition Phases*, AFSCM 375-1, Department of the Air Force, Washington, DC, 1962.
10. Hass, A.M.J., *Configuration Management Principles and Practice*, Addison-Wesley, Boston, 2003.
11. *Engineering Design Handbook: Value Engineering*, AMCP 706-104, Department of Defense, Washington, DC, 1971.
12. Demarle, D.J. and Shillito, M.L., Value engineering, in *Handbook of Industrial Engineering*, Selvendy, G., Ed., John Wiley & Sons, New York, 1982, pp. 7.3.1–7.3.20.
13. Younker, D.L., *Value Engineering: Analysis and Methodology*, Marcel Dekker, New York, 2003.
14. Riggs, J.L. and Inoue, M.S., *Introduction to Operations Research and Management Science: A General Systems Approach*, McGraw-Hill, New York, 1975.
15. Lomax, P.A., *Network Analysis: Application to the Building Industry*, The English Universities Press, London, 1969.
16. Antill, J.M., *Critical Path Methods in Construction Practice*, John Wiley & Sons, New York, 1982.
17. Thomas, E.F., Pitfalls in reliability program management, *Proceedings of the Annual Reliability and Maintainability Symposium*, 369–373, 1976.

6 Mechanical and Human Reliability

6.1 INTRODUCTION

Usually, the concept of constant failure rate (i.e., exponentially distributed times to failure) is used to evaluate the reliability of electronic components. This concept may or may not be applicable to mechanical parts. The history of mechanical reliability may be traced back to World War II and the development of V1 and V2 rockets in Germany. However, it was not until the mid-1960s when mechanical reliability received serious attention because of the loss of two spacecrafts (i.e., Syncom I and Mariner III) because of mechanical failures [1]. Consequently, the National Aeronautics and Space Administration (NASA) initiated many projects to improve reliability of mechanical items.

Today, mechanical reliability has become an important element of the reliability engineering field. A comprehensive list of publications on mechanical reliability is available in Reference 2.

Many times engineering systems fail because of human errors rather than because of hardware or software failures. The history of human reliability may be traced back to the late 1950s when H.L. Williams pointed out that human-element reliability must be included in the overall system reliability prediction; otherwise, such a prediction would not be realistic [3].

Many people have contributed to human reliability. The first book on the subject appeared in 1986 [4]. A comprehensive list of publications on human reliability is available in Reference 5. This chapter presents important aspects of mechanical reliability and human reliability.

6.2 GENERAL MECHANICAL FAILURE CAUSES AND MODES

Various people have studied the causes of mechanical failures and have identified them as follows [6]:

- Poor or defective design
- Manufacturing defect
- Incorrect application
- Wrong installation
- Wear-out
- Failure of other parts or components
- Gradual deterioration in performance

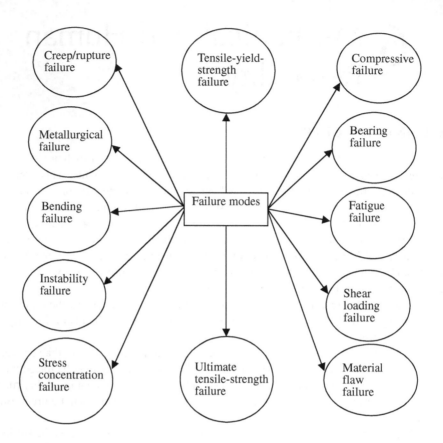

FIGURE 6.1 Mechanical failure modes.

Many different types of failure modes are associated with mechanical parts or items. Some of these failure modes are shown in Figure 6.1 [7–9]. These are fatigue failure, creep or rupture failure, bending failure, metallurgical failure, instability failure, shear loading failure, material flaw failure, compressive failure, bearing failure, stress concentration failure, ultimate tensile-strength failure, and tensile-yield-strength failure.

Fatigue failure occurs because of repeated loading or unloading (or partial unloading) of an item or part. Its occurrence can be prevented by selecting appropriate materials for a specific application. For example, under cycle loading, steel outlasts aluminum. In the case of creep or rupture failure, material stretches (i.e., creeps) when the load is maintained on a continuous basis, and normally it ultimately terminates in a rupture. Also, creep accelerates with elevated temperatures.

Bending failure occurs when one outer surface is in compression and the other outer surface is in tension. An example of the bending failure is the tensile rupture of the outer material. Metallurgical failure is also known as a material failure. This type of failure is the result of extreme oxidation or operation in a corrosive environment. The occurrence of metallurgical failures is accelerated by environmental conditions such as heat, erosion, nuclear radiation, and corrosive media.

The instability failure is confined to structural members such as beams and columns, in particular, those manufactured using thin material where the loading is normally in compression. However, this type of failure may also occur because of torsion or by combined loading (i.e., compression and bending). The shear loading failure occurs when shear stress becomes greater than the strength of the material when applying high shear or torsion loads.

Material flaw failure occurs because of factors such as weld defects, poor quality assurance, small cracks and flaws, and fatigue cracks. Compressive failure causes permanent deformation, rupturing, or cracking and is similar to tensile failures except under compressive loads.

Bearing failure usually occurs because of a cylindrical surface bearing on either a flat or a concave surface like roller bearings in a race and is similar in nature to compressive failure. The stress concentration failure occurs under the conditions of uneven stress flow through a mechanical design.

Ultimate tensile-strength failure occurs when the ultimate tensile strength is less than the applied stress and leads to a complete failure of the structure at a cross-sectional point. Tensile-yield-strength failure occurs under tension and, more specifically, when the applied stress is greater than the material yield strength.

6.3 SAFETY FACTORS

Various safety factors are used during design to ensure reliability of mechanical items. They are arbitrary multipliers. They can be quite useful to provide satisfactory design if they are established with utmost care, using considerable past experiences and data.

Three safety factors are presented below.

6.3.1 SAFETY FACTOR 1

This is defined by [10]

$$SF = \frac{SL_m}{SL_n} \tag{6.1}$$

where SF is the safety factor, SL_n is the normal service load, and SL_m is the maximum safe load.

This safety factor is considered quite good, particularly when the loads are distributed.

6.3.2 SAFETY FACTOR 2

This is defined by [8, 9]

$$SF = \frac{S_m}{WS_m} \tag{6.2}$$

where SF is the safety factor, WS_m is the maximum allowable working stress, and S_m is the strength of the material.

The value of this specific safety factor is always greater than unity. More specifically, when its value is less than unity, it simply means that the item under consideration will fail because its strength is less than the applied stress.

The standard deviation of this safety factor is given by [8, 9]

$$\sigma = \left[\left(\sigma_{th} / WS_m \right)^2 + \left\{ S_m / \left(WS_m \right)^2 \right\} \sigma_{st}^2 \right]^{1/2} \tag{6.3}$$

where
σ is the safety factor standard deviation, σ_{st} is the stress standard deviation, and σ_{th} is the strength standard deviation.

6.3.3 SAFETY FACTOR 3

This is defined by [11]

$$SF = \frac{STH_m}{ST_m} \geq 1 \tag{6.4}$$

where
SF is the safety factor, STH_m is the mean failure-governing strength, and ST_m is the mean failure-governing stress.

Generally, for normally distributed stress and strength, this safety factor is a good measure. However, for large variations in stress or strength, this safety factor becomes meaningless because of the positive failure rate.

All in all, for selecting an appropriate value of the safety factor, careful consideration must be given to factors such as cost, failure consequence, uncertainty of material strength, load uncertainty, and the degree of uncertainty in relating applied stress to strength [12].

6.4 STRESS–STRENTH INTERFERENCE THEORY MODELING

When the probability density functions of an item's stress and strength are known, its reliability may be determined analytically. The approach used for determining reliability is known as stress–strength interference theory modeling. Reliability is simply the probability that the failure-governing stress will not exceed the failure-governing strength. Mathematically, it is expressed as follows [13, 14]:

$$R = P\left(x < y \right) = P\left(y > x \right) \tag{6.5}$$

where R is the item reliability, x is the stress random variable, y is the strength random variable, and P is the probability.

With the aid of Reference 14, we rewrite Equation 6.5 as follows:

$$R = \int_{-\infty}^{\infty} f(x) \left[\int_{x}^{\infty} f(y)\,dy \right] dx \tag{6.6}$$

where $f(x)$ is the stress probability density function and $f(y)$ is the strength probability density function.

In the published literature Equation 6.6 is also written in the following three forms [14]:

$$R = \int_{-\infty}^{\infty} f(y) \left[1 - \int_{y}^{\infty} f(x)\,dx \right] dy \tag{6.7}$$

$$R = \int_{-\infty}^{\infty} f(x) \left[1 - \int_{-\infty}^{x} f(y)\,dy \right] dx \tag{6.8}$$

$$R = \int_{-\infty}^{\infty} f(y) \left[\int_{-\infty}^{y} f(x)\,dx \right] dy \tag{6.9}$$

The application of Equation 6.6 is demonstrated by developing the following two stress–strength models when the stress and strength probability density functions of an item are known.

6.4.1 Model 1

This model assumes that the stress and strength associated with an item are exponentially distributed. Thus, we have

$$f(x) = \theta e^{-\theta x}, \quad 0 \le x < \infty \tag{6.10}$$

and

$$f(y) = \lambda e^{-\lambda y}, \quad 0 \le y < \infty \tag{6.11}$$

where
θ and λ are the reciprocals of the mean values of stress and strength, respectively.

By substituting Equation 6.10 and Equation 6.11 into Equation 6.6, we get

$$\begin{aligned}
R &= \int_{0}^{\infty} \theta e^{-\theta x} \left[\int_{x}^{\infty} \lambda e^{-\lambda y}\,dy \right] dx \\
&= \int_{0}^{\infty} \theta \, e^{-(\theta + \lambda)x}\,dx \\
&= \frac{\theta}{\theta + \lambda}
\end{aligned} \tag{6.12}$$

For $\theta = \frac{1}{\bar{x}}$ and $\lambda = \frac{1}{\bar{y}}$, Equation 6.12 becomes

$$R = \frac{\bar{y}}{\bar{y} + \bar{x}} \tag{6.13}$$

where
$\bar{x}$ and $\bar{y}$ are the mean stress and strength, respectively.

Example 6.1

An item's stress and strength are exponentially distributed with mean values of 4,000 and 28,000 psi, respectively. Calculate the item reliability.

By substituting the specified data values into Equation 6.13, we get

$$R = \frac{28,000}{28,000 + 4,000}$$
$$= 0.875$$

Thus, the item reliability is 0.875.

6.4.2 MODEL 2

This model assumes that the stress and strength of an item follow exponential and normal distributions, respectively. Thus, we have

$$f(x) = \theta e^{-\theta x}, \quad x > 0 \tag{6.14}$$

$$f(y) = \frac{1}{\sigma_y \sqrt{2\Pi}} \exp\left[-\frac{1}{2}\left(\frac{y - \mu_y}{\sigma_y}\right)^2\right], \quad -\infty < y < \infty \tag{6.15}$$

where μ_y is the mean strength, θ is the reciprocal of the mean stress, and σ_y is the strength standard deviation.

By substituting Equation 6.14 and Equation 6.15 into Equation 6.9, we get

$$R = \int_{-\infty}^{\infty} \frac{1}{\sigma_y \sqrt{2\Pi}} \exp\left[-\frac{1}{2}\left(\frac{y - \mu_y}{\sigma_y}\right)^2\right] dy \left[\int_0^y \theta e^{-\theta x} dx\right]$$

$$= 1 - \int_{-\infty}^{\infty} \frac{1}{\sigma_y \sqrt{2\Pi}} \exp\left\{-\left[\frac{1}{2}\left(\frac{y - \mu_y}{\sigma_y}\right)^2 + \theta y\right]\right\} dy \tag{6.16}$$

Since

$$\frac{1}{2}\left(\frac{y-\mu_y}{\sigma_y}\right)^2 + \theta\, y = \frac{2\mu_y\,\theta\,\sigma_y^2 + \left(\sigma_y^2\,\theta - \mu_y + y\right)^2 - \sigma_y^4\,\theta^2}{2\sigma_y^2} \tag{6.17}$$

Equation 6.16 is rewritten to the following form:

$$R = 1 - \exp\left[-\frac{1}{2}\left(2\mu_y\,\theta - \sigma_y^2\,\theta^2\right)\right] M \tag{6.18}$$

where

$$M = \int_{-\infty}^{\infty} \frac{1}{\sigma_y\sqrt{2\Pi}}\,\exp\left[-\frac{1}{2}\left(\frac{\sigma_y^2\,\theta - \mu_y + y}{\sigma_y}\right)^2\right] dy \tag{6.19}$$

Since $M = 1$ [15], Equation 6.18 reduces to

$$R = 1 - \exp\left[-\frac{1}{2}\left(2\mu_y\,\theta - \sigma_y^2\,\theta^2\right)\right] \tag{6.20}$$

Example 6.2

Assume that the stress and strength associated with an item are described by exponential and normal distributions, respectively. The mean value of the stress is 6,000 psi, and the mean and standard deviation of the strength are 25,000 and 2,000 psi, respectively. Compute the item reliability.

By substituting the given data values into Equation 6.20, we get

$$R = 1 - \exp\left[-\frac{1}{2}\left\{\frac{2(25000)}{(6000)} - \frac{(2000)^2}{(6000)^2}\right\}\right]$$

$$= 0.9836$$

Thus, the item's reliability is 0.9836.

6.5 GRAPHICAL METHOD FOR ESTIMATING MECHANICAL ITEM'S RELIABILITY

This method uses Mellin transforms in estimating the reliability of a mechanical item [16]. The approach is extremely useful when its associated stress and strength probability distributions cannot be assumed but there is a sufficient amount of empirical data. However, this method can also be used when an item's stress and strength probability distributions are known.

The Mellin transforms for Equation 6.6 are

$$D = \int_x^\infty f(y) \, dy$$
$$= 1 - F_1(x) \tag{6.21}$$

and

$$C = \int_0^x f(x) \, dx = F_2(x) \tag{6.22}$$

where
$F_1(x)$ and $F_2(x)$ are the cumulative distribution functions.
Differentiating Equation 6.22 with respect to x, we get

$$\frac{dC}{dx} = f(x) \tag{6.23}$$

Rearranging Equation 6.23 results in

$$dC = f(x).dx \tag{6.24}$$

It is quite obvious from Equation 6.22 that C takes values of 0 to 1 (i.e., at $x = 0$, $C = 0$ and at $x = \infty$, $C = 1$).
By inserting Equation 6.21 and Equation 6.24 into Equation 6.6, we get

$$R = \int_0^1 D \, dC \tag{6.25}$$

Equation 6.25 indicates that the reliability of the item is given by the area under the C versus D plot. This area can be calculated by using Simpson's rule, presented below [17]:

$$\int_i^n f(z) \, dz \cong \frac{n-i}{3m} \left(X_0 + 4X_1 + 2X_2 + 4X_3 + ... + 2X_{m-2} + 4X_{m-1} + X_m \right) \tag{6.26}$$

where $f(z)$ is a function of z over interval (i, n), that is, $i \leq z \leq n$, m is the even number of equal subdivided parts of intervals (i, n), and $n - i/3m = w$ is the subdivided part width.
At $i = z_0, z_1, z_2, ..., z_m = n$; the corresponding values of X are $X_0 = f(z_0)$, $X_1 = f(z_1)$, $X_2 = f(z_2)$, ..., $X_m = f(z_m)$.

Example 6.3

With the aid of the graphical method, estimate the item reliability in Example 6.1.
Inserting Equation 6.10 and the specified data into Equation 6.22 yields

$$C = \int_0^x \frac{1}{4,000} e^{-\left(\frac{1}{4,000}\right)x} dx$$

$$= 1 - e^{-\left(\frac{1}{4,000}\right)x}$$

(6.27)

By substituting Equation 6.11 and the specified relevant data into Equation 6.21,
we get

$$D = \int_x^\infty \frac{1}{28,000} e^{-\left(\frac{1}{28,000}\right)y} dy$$

$$= e^{-\left(\frac{1}{28,000}\right)x}$$

(6.28)

For the assumed values of stress x and using Equation 6.27 and Equation 6.28,
the values of C and D, are tabulated in Table 6.1. Figure 6.2 shows the plot of the

TABLE 6.1
Calculated Values of C and D for the Assumed Values of Stress x

x (psi)	C	D
0	0.00000	1.00000
500	0.11750	0.98230
1,000	0.22120	0.96492
1,500	0.31271	0.94784
2,000	0.39347	0.93106
2,500	0.46474	0.91458
3,000	0.52763	0.89840
3,500	0.58314	0.88250
4,000	0.63212	0.86688
4,500	0.67535	0.85154
5,000	0.71350	0.83646
6,000	0.77687	0.80712
7,000	0.82623	0.77880
8,000	0.86466	0.75148
9,000	0.89460	0.72511
10,000	0.91792	0.69967
15,000	0.97648	0.58525
20,000	0.99326	0.48954
25,000	0.99807	0.40948
30,000	0.99945	0.34252
Infinity	1.00000	0.00000

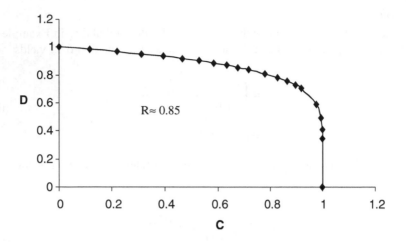

FIGURE 6.2 Plot of C versus D.

C and D values in Table 6.1. The total area under the Figure 6.2 curve is calculated using Equation 6.26 as follows:

$$R \cong \frac{(1-0)}{3(4)}\left[X_0 + 4X_1 + 2X_2 + 4X_3 + X_4\right]$$

$$\cong \frac{1}{12}\left[1 + 4(0.98) + 2(0.96) + 4(0.84) + 0\right]$$

$$R \cong 0.85$$

Thus, the item reliability is approximately 0.85. This value is close to but lower than the one obtained using the analytical approach in Example 6.1 (i.e., 0.875).

6.6 HUMAN ERROR OCCURRENCE FACTS AND FIGURES

Many studies concerning the occurrence of human errors have been conducted. Results of some of their findings are as follows:

- A study of 135 vessel failures that occurred during the period from 1926 to 1988 revealed that about 25% of the failures were caused by humans [18].
- Over 90% of the documented air traffic control system errors were caused by human operators [19].
- A study of 23,000 defects in the production of nuclear parts discovered that around 82% of the defects were caused by humans [20].

- Each year about 100,000 Americans die because of human errors in health care, and the annual financial impact on the U.S. economy is estimated to be somewhere between $17 billion and $29 billion [21].
- Around 60% of all medical device–related deaths and injuries reported through the Center for Devices and Radiological Health (CDRH) of the Food and Drug Administration (FDA) were caused by human errors [22].
- A total of 401 human errors occurred in U.S. commercial light-water nuclear reactors during the period from June 1, 1973 to June 30, 1975 [23].

6.7 HUMAN ERROR CATEGORIES AND CAUSES

Human errors may be classified into many categories as shown in Figure 6.3 [24–26].

Operator errors are the result of operator mistakes, and the causes of their occurrence include poor environment, complex tasks, lack of proper procedures, operator carelessness, and poor personnel selection and training. Maintenance errors occur in field environments because of oversights by maintenance personnel. Some examples of maintenance errors are repairing a failed item incorrectly, calibrating equipment incorrectly, and applying the wrong grease at appropriate points on the equipment.

Assembly errors are the result of human mistakes during product assembly. Some of the causes of assembly errors are poor illumination, poor blueprints and other related material, poorly designed work layout, and poor communication of related information. Installation errors occur for various reasons including failure to install equipment or items per the manufacturer's specification and using the incorrect installation instructions or blueprints.

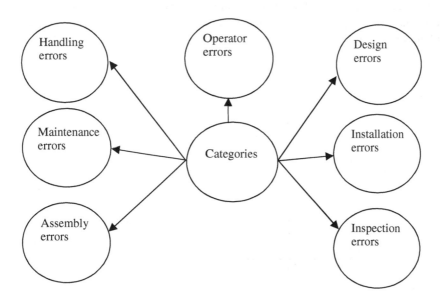

FIGURE 6.3 Human error categories.

Design errors are the result of inadequate design. Some of the causes of their occurrence are failure to ensure the effectiveness of person–machine interactions, failure to implement human needs in the design, and assigning inappropriate functions to humans. Inspection errors are the result of less than 100% accuracy of inspection personnel. One typical example of inspection errors is accepting and rejecting out-of-tolerance and in-tolerance components and items, respectively. Handling errors occur because of improper transportation or storage facilities.

There are many causes of human errors. Some of the common ones are poor training or skills of personnel, inadequate work tools, poor motivation of personnel, poorly written product and equipment operating and maintenance procedures, complex tasks, poor work layout, poor equipment and product design, and poor job environment (i.e., poor lighting, crowded work space, high noise level, high or low temperature, etc.) [25–26].

6.8 HUMAN STRESS–PERFORMANCE EFFECTIVENESS AND STRESS FACTORS

Researchers have studied the relationship between human performance effectiveness and stress. Figure 6.4 shows the resulting curve of their effort [27–28]. This curve shows that a moderate level of stress is necessary for increasing the effectiveness of human performance to its maximum. The moderate level may simply be interpreted as high enough stress to keep the individual alert.

At a very low stress, the task becomes dull and unchallenging; therefore, most people will not perform effectively and their performance will not be at the optimum level. When the stress passes its moderate level, the effectiveness of the human performance starts to decline. This decline is mainly due to factors such as worry, fear, and other types of psychological stress. At the highest stress level, human reliability is at its lowest level.

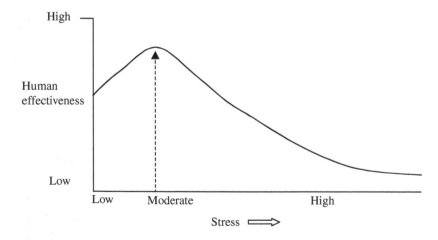

FIGURE 6.4 A hypothetical human performance effectiveness–stress curve.

Researchers have noted a large number of factors that can increase the stress on humans and in turn decrease their reliability in work and other environments. Some of these factors are dissatisfaction with the job, possibility of work layoff, inadequate expertise to perform the task, excessive demands of superiors, current job being below ability and experience, tasks being performed under extremely tight time schedules, serious financial problems, low chance of promotion, health problems, difficulties with spouse or children, working with people who have unpredictable temperaments [27].

6.9 HUMAN PERFORMANCE RELIABILITY IN CONTINUOUS TIME AND MEAN TIME TO HUMAN ERROR MEASURE

Humans perform various types of time-continuous tasks. Some examples of these tasks are scope monitoring, aircraft maneuvering, and missile countdown. The following equation, developed the same way as the general reliability function, can be used to calculate performance reliability [25]:

$$R_{hp}(t) = e^{\int_0^t \lambda_h(t)\,dt} \tag{6.29}$$

where $R_{hp}(t)$ is the human reliability at time t and $\lambda_h(t)$ is the time-dependent human error rate.

Equation 6.29 is known as the general human performance reliability function. More specifically, it can be used to calculate human reliability at time t when time to human error follows any known probability distribution.

A general expression for mean time to human error is given by [25]

$$MTTHE = \int_0^\infty R_{hp}(t)\,dt$$
$$= \int_0^\infty \exp\left[-\int_0^t \lambda_h(t)\,dt\right]dt \tag{6.30}$$

where $MTTHE$ is the mean time to human error.

Equation 6.30 can be used to obtain the $MTTHE$ when times to human error follow any probability distribution. The application of Equation 6.29 and Equation 6.30 is demonstrated through the following example.

Example 6.4

A person is performing a time-continuous task and his or her times to error are exponentially distributed. Calculate the person's mean time to error and reliability for an 8-hour mission if his or her error rate is 0.008 errors per hour.

Thus, in this case we have

$$\lambda_h(t) = \lambda_h = 0.008 \; errors\,/\,hour$$

and

$$t = 8 \text{ hours}$$

where λ_h is the person's constant error rate.

By substituting the above values into Equation 6.29, we get

$$R_{hp}(8) = e^{-\int_0^8 (0.008)\, dt}$$

$$= e^{-(0.008)(8)}$$

$$= 0.9380$$

Similarly, using the specified data values in Equation 6.30 yields

$$MTTHE = \int_0^\infty \exp\left[-\int_0^8 (0.008)\, dt\right] dt$$

$$= \frac{1}{(0.008)}$$

$$= 125 \ hours$$

Thus, the person's mean time to error and reliability are 125 hours and 0.9380, respectively.

6.10 METHODS FOR PERFORMING HUMAN RELIABILITY ANALYSIS

Many methods can be used to perform various types of human reliability analysis [25]. Each has its advantages and disadvantages. Two of these methods are presented below.

6.10.1 FAULT TREE ANALYSIS

This method is described in detail in Chapter 8 and it can also be used to conduct human reliability analysis [25,29,30]. Its application in human reliability work is demonstrated through the following two examples.

Example 6.5

Assume that a person is required to do a certain operation-related job: job A. The job is composed of three independent tasks B, C, and D. If any one of these three tasks is performed incorrectly, job A will not be accomplished successfully.

Task B is composed of two subtasks b_1 and b_2. For the successful performance of task B, only one of these subtasks needs to be performed correctly. Subtask b_1 is composed of three independent steps i, j, and k. All these three steps must be performed correctly for the successful completion of subtask b_1.

Develop a fault tree for the top event: job A will not be accomplished correctly by the person.

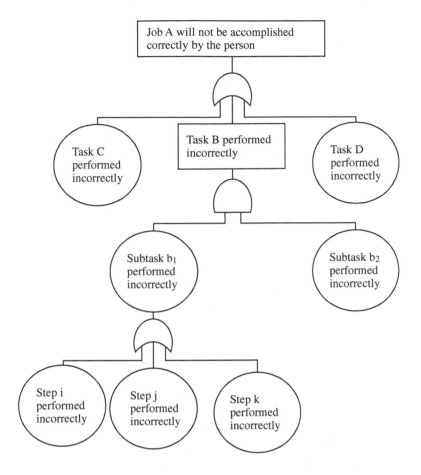

FIGURE 6.5 Fault tree for the top event: job A will not be accomplished correctly.

Using the Chapter 8 fault tree symbols, the fault tree for the example is shown in Figure 6.5.

Example 6.6

Assume that in Figure 6.5 the probability of occurrence of basic events C, D, B_2, i, j, and k is 0.07. Calculate the probability of occurrence of the top event (i.e., job A will not be accomplished correctly) if all the events occur independently.

The probability of performing subtask b_1 incorrectly is given by [31]

$$P(b_1)=1-\{1-P(i)\}\{1-P(j)\}\{1-P(k)\}$$
$$=1-\{1-0.07\}\{1-0.07\}\{1-0.07\}$$
$$=0.1956$$

where $P(i)$ is the probability of performing step i incorrectly, $P(j)$ is the probability of performing step j incorrectly, and $P(k)$ is the probability of performing step k incorrectly.

The probability of performing task B incorrectly is given by [31]

$$P(B) = P(b_1)P(b_2)$$
$$= (0.1956)(0.07)$$
$$= 0.0137$$

where $P(b_1)$ is the probability of performing subtask b_i incorrectly for i = 1, 2. The probability of accomplishing job A incorrectly is

$$P(A) = 1 - \{1 - P(C)\}\{1 - P(B)\}\{1 - P(D)\}$$
$$= 1 - \{1 - 0.07\}\{1 - 0.0137\}\{1 - 0.07\}$$
$$= 0.1469$$

Figure 6.6 shows the Figure 6.5 fault tree with the above calculated and specified fault event occurrence probability values. The probability of occurrence of the top event (job A will not be accomplished correctly) is 0.1469.

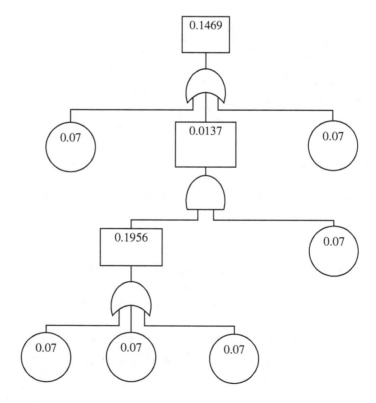

FIGURE 6.6 Fault tree with specified and calculated fault event occurrence probability values.

6.10.2 Markov Method

This is a widely used method in reliability engineering and it can also be used to conduct time-continuous human reliability analysis. The method is described in Chapter 4 and is based on the following assumptions [32]:

- All occurrences are independent of each other.
- The transitional probability from one state to another in finite time interval Δt is given by $\lambda_{he} \Delta t$. The parameter λ_{he} is the constant human error rate.
- The probability of more than one transitional occurrences in finite time Δt is negligible [i.e., $(\lambda_{he} \Delta t)(\lambda_{he} \Delta t) \to 0$].

The application of this method in human reliability work is demonstrated through the following example:

Example 6.7

Assume that a person is performing a time-continuous task and his or her constant error rate is λ_{he}. The state–space diagram of the person performing the task is shown in Figure 6.7. Develop probability expressions for the person performing the task successfully and unsuccessfully (i.e., the person commits an error) at time t by using the Markov method. Also, develop an expression for the mean time to error committed by the person.

In Figure 6.7 numerals denote corresponding states (i.e., 0: person performing his or her task normally or successfully, 1: person committed an error).

The following symbols are associated with Figure 6.7:

- λ_{he} is the person's constant error rate (i.e., constant human error rate).
- $P_0(t)$ is the probability that the person is performing the task normally at time t.
- $P_1(t)$ is the probability that the person has committed an error at time t.

Using the Markov method, we write down the following two equations for Figure 6.7:

$$P_0(t+\Delta t) = P_0(t)(1 - \lambda_{he} \Delta t) \tag{6.31}$$

$$P_1(t+\Delta t) = (\lambda_{he} \Delta t) P_0(t) + P_1(t) \tag{6.32}$$

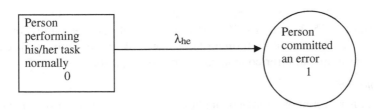

FIGURE 6.7 State–space diagram for a person performing a time-continuous task.

where $(1-\lambda_{he}\,\Delta t)$ is the probability of the occurrence of no error in finite time interval Δt, $P_0\,(t+\Delta t)$ is the probability that the person is performing the task successfully or normally at time $(t+\Delta t)$, and $P_1\,(t+\Delta t)$ is the probability that the person has committed an error at time $(t+\Delta t)$.

By rearranging Equation 6.31 and Equation 6.32 and taking the limits, we get

$$\frac{d\,P_0\,(t)}{dt}=-\lambda_{he}\,P_0\,(t) \tag{6.33}$$

$$\frac{d\,P_1\,(t)}{dt}=\lambda_{he}\,P_0\,(t) \tag{6.34}$$

At time $t=0$, $P_0\,(0)=1$ and $P_1\,(0)=0$.
Solving Equation 6.33 and Equation 6.34 using Laplace transforms [4] results in

$$P_0\,(t)=e^{-\lambda_{he}t} \tag{6.35}$$

$$P_1\,(t)=1-e^{-\lambda_{he}t} \tag{6.36}$$

Thus, the person's reliability is given by

$$R_p\,(t)=P_0\,(t)=e^{-\lambda_{he}t} \tag{6.37}$$

where $R_p\,(t)$ is the person's reliability at time t.
Mean time to error committed by the person is given by [31]

$$\begin{aligned}MTTECP&=\int_0^\infty R_p\,(t)\,dt\\ &=\int_0^\infty e^{-\lambda_{he}t}dt\\ &=\frac{1}{\lambda_{he}}\end{aligned} \tag{6.38}$$

where $MTTECP$ is the mean time to error committed by the person.

Equation 6.35, Equation 6.36, and Equation 6.38 are the expressions for the probability of the person performing his or her task successfully, unsuccessfully, and the mean time to error committed by the person, respectively.

Example 6.8

A person is performing a time-continuous task and his or her constant error rate is 0.001 errors per hour. Calculate the person's reliability during a 10-hour mission and mean time to error.

By inserting the specified data into Equation 6.37 and Equation 6.38 we get

$$R_p(10) = e^{-(0.001)(10)}$$
$$= 0.99$$

and

$$MTTECP = \frac{1}{0.001}$$
$$= 1,000 \ hours$$

Thus, the person's reliability and mean time to error are 99% and 1,000 hours, respectively.

6.11 PROBLEMS

1. Write an essay on the history of mechanical reliability.
2. List at least 12 mechanical failure modes.
3. What are the general causes for the occurrence of mechanical failures?
4. Write down at least two distinct definitions of the safety factor.
5. Prove that Equation 6.6, through Equation 6.9 are identical.
6. Discuss the following types of human errors:
   • Design errors
   • Maintenance errors
   • Operator errors
7. Discuss factors that increase stress on humans.
8. Prove that for $\lambda_h(t) = \lambda_{he}$ Equation 6.29 and Equation 6.30 give same results as Equation 6.37 and Equation 6.38, respectively.
9. Assume that the stress and strength associated with an item are described by exponential and normal distributions, respectively. The mean value of the stress is 7,000 psi, and the mean and standard deviation of the strength are 20,000 psi and 1,000 psi, respectively. Calculate the item reliability.
10. Discuss the human performance effectiveness versus stress curve.

REFERENCES

1. Redler, W.M., Mechanical reliability research in the National Aeronautics and Space Administration, *Proceedings of the Reliability and Maintainability Conference*, 763–768, 1966.
2. Dhillon, B.S., Reliability and Quality Control: Bibliography on General and Specialized Areas, Beta Publishers, Gloucester, Ontario, Canada, 1992.
3. Williams, H.L., Reliability evaluation of the human component in man-machine systems, *Electrical Manufacturing*, April, 78–82, 1958.

4. Dhillon, B.S., Human Reliability: With Human Factors, Pergamon Press, New York, 1986.
5. Dhillon, B.S. and Yang, N., Human reliability: a literature survey and review, *Microelectronics and Reliability*, 34, 803–810, 1994.
6. Lipson, C., Analysis and Prevention of Mechanical Failures, Course Notes No. 8007, University of Michigan, Ann Arbor, MI, 1980.
7. Collins, J.A., Failure of Materials in Mechanical Design, John Wiley & Sons, New York, 1981.
8. Doyle, R.L., Mechanical-System Reliability, Tutorial Notes, Annual Reliability and Maintainability Symposium, 1992.
9. Grant Ireson, W., Coombs, C.F., and Moss, R.Y., Handbook of Reliability Engineering and Management, McGraw-Hill, New York, 1996.
10. Phelan, M.R., Fundamentals of Machine Design, McGraw-Hill, New York, 1962.
11. Bompass-Smith, J.H., Mechanical Survival: The Use of Reliability Data, McGraw-Hill, London, 1973.
12. Juvinall, R.C., Fundamentals of Machine Component Design, John Wiley & Sons, New York, 1983.
13. Kececioglu, D., Reliability analysis of mechanical components and systems, *Nuclear Engineering and Design*, 19, 249–290, 1972.
14. Dhillon, B.S., Mechanical Reliability: Theory, Models, and Applications, American Institute of Aeronautics and Astronautics, Washington, DC, 1988.
15. Kececioglu, D. and Li, D., Exact solutions for the prediction of the reliability of mechanical components and structural members, in *Proceedings of the Failure Prevention Reliability Conference*, American Society of Mechanical Engineers (ASME), New York, 1985, pp. 115–122.
16. *Quality Assurance (Reliability Handbook)*, AMCP 702-3, Department of Defense, Washington, DC, 1968.
17. Spiegel, M.R., *Mathematical Handbook of Formulas and Tables*, McGraw-Hill, New York, 1968.
18. *Organizational Management and Human Factors in Quantitative Risk Assessment*, Report No. 33/1992 (Report 1), British Health and Safety Executive (HSE), London, 1992.
19. Kenney, G.C., Spahn, M.J., and Amato, R.A., *The Human Element in Air Traffic Control: Observations and Analysis of Controllers and Supervisors in Providing Air Traffic Control Separation Services*, Report No. MTR-7655, METREK Div., MITRE Corporation, 1977.
20. Rook, L.W., *Reduction of Human Error in Industrial Production*, Report No. SCTM 93–63(14), Sandia National Laboratories, Albuquerque, NM, 1962.
21. Kohn, L.T., Corrigan, J.M., and Donaldson, M.S., Eds., *To Err Is Human: Building a Safer Health System*, Institute of Medicine Report, National Academy Press, Washington, DC, 1999.
22. Bogner, M.S., Medical devices: a new frontier for human factors, *CSERIAC Gateway*, 4(1), 12–14, 1993.
23. Joos, D.W., Sabri, Z.A., and Husseiny, A.A., Analysis of gross error rates in operation of commercial nuclear power stations, *Nuclear Engineering Design*, 52, 265–300, 1979.
24. Cooper, J.J., Human-initiated failures and malfunction reporting, *IRE Transactions on Human Factors*, 10, 104–109, 1961.
25. Dhillon, B.S., *Human Reliability: With Human Factors*, Pergamon Press, New York, 1986.

26. Meister, D., The problem of human-initiated failures, *Proceedings of the 8th National Symposium on Reliability and Quality Control*, 1962, pp. 234–239.
27. Beech, H.R., Burns, L.E., and Sheffield, B.F., *A Behavioral Approach to the Management of Stress*, John Wiley & Sons, New York, 1982.
28. Hagen, E.W., Human reliability analysis, *Nuclear Safety*, 17, 315–326, 1976.
29. *Fault Tree Handbook*, NUREG-0492, U.S. Nuclear Regulatory Commission, Washington, DC, 1981.
30. Dhillon, B.S. and Singh, C., *Engineering Reliability: New Techniques and Applications*, John Wiley and Sons, New York, 1981.
31. Dhillon, B.S., *Design Reliability: Fundamentals and Applications*, CRC Press, Boca Raton, FL, 1999.
32. Shooman, M.L., *Probabilistic Reliability: An Engineering Approach*, McGraw-Hill, New York, 1968.

7 Introduction to Engineering Maintainability

7.1 NEED FOR MAINTAINABILITY

The need for maintainability is becoming more important than ever before because of the alarmingly high operating and support costs of equipment and systems. For example, each year the United States industry spends over $300 billion on plant maintenance and operations [1]. Furthermore, the annual cost of maintaining a military jet aircraft is around $1.6 million; approximately 11% of the total operating cost for an aircraft is spent on maintenance activities [2].

Some of the objectives of applying maintainability engineering principles are to reduce projected maintenance time and costs, to determine labor-hours and other related resources required for performing the projected maintenance, and to use maintainability data to estimate equipment availability or unavailability.

When maintainability engineering principles are applied successfully to any product, results such as reduction in product downtime, efficient restoration of the product to its operating state, and maximum operational readiness of the product can be expected [3].

7.2 ENGINEERING MAINTAINABILITY VERSUS ENGINEERING MAINTENANCE

As maintainability and maintenance are closely interrelated, many people find it difficult to make a clear distinction between them. Maintainability refers to measures or steps taken during the product design phase to include features that will increase ease of maintenance and ensure that the product will have minimum downtime and life cycle support costs when used in field environments [3]. In contrast, maintenance refers to measures taken by the product users for keeping it in operational state or repairing it to operational state [1,4].

More simply, maintainability is a design parameter intended to minimize equipment repair time, whereas maintenance is the act of servicing and repairing equipment [5].

The responsibility of the maintenance engineers is to ensure that product or equipment design and development requirements reflect the maintenance needs of users. Thus, they are concerned with factors such as the environment in which the product will be operated and maintained; product and system mission, operational, and support profiles; and the levels and types of maintenance required. Product maintainability design requirements are determined by various processes including

TABLE 7.1
Specific General Principles of Maintainability and Reliability

No.	Specific General Principle: Maintainability	Specific General Principle: Reliability
1	Reduce life cycle maintenance costs	Maximize the use of standard parts
2	Reduce the amount, frequency, and complexity of required maintenance tasks	Use fewer components for performing multiple functions
3	Reduce mean time to repair (MTTR)	Design for simplicity
4	Determine the extent of preventive maintenance to be performed	Provide adequate safety factors between strength and peak stress values
5	Provide for maximum interchange ability	Provide fail-safe designs
6	Reduce the amount of supply supports required	Provide redundancy when required
7	Reduce or eliminate the need for maintenance	Minimize stress on components and parts
8	Consider benefits of modular replacement versus part repair or throwaway design	Use parts and components with proven reliability

the analysis of maintenance tasks and requirements, the determination of maintenance resource needs, the development of maintenance concepts, and maintenance
engineering analysis [6].

7.3 MAINTAINABILITY* VERSUS RELIABILITY

Maintainability is a built-in design and installation characteristic that provides the
resulting equipment or product with an inherent ability to be maintained, leading to
factors such as better mission availability and lower maintenance cost, required tools
and equipment, required skill levels, and required man-hours.

In contrast, reliability is a design characteristic that leads to durability of the
equipment as it performs its assigned function according to a specified condition and
time period. It is accomplished through actions such as choosing optimum engineering
principles, testing, controlling processes, and satisfactory component sizing.

Some of the important specific general principles of maintainability and reliability
are presented in Table 7.1 [7].

7.4 MAINTAINABILITY FUNCTIONS

Just like in any other area of engineering, probability distributions play an important
role in maintainability engineering. They are used to represent repair times of
equipment, systems, and parts. After identification of the repair distribution, the
corresponding maintainability function may be obtained. The maintainability function

---

* Some of the material concerning maintainability in this section may overlap with the material presented
in the previous section.

is concerned with predicting the probability that a repair, beginning at time $t = 0$, will be completed in a time t.

Mathematically, the maintainability function is defined by [3]

$$M(t) = \int_0^t f_{dr}(t)\, dt \qquad (7.1)$$

where t is time, $f_{dr}(t)$ is the probability density function of the repair time, and $M(t)$ is the maintainability function.

Maintainability functions for various probability distributions are obtained below [3,6,8–10].

7.4.1 MAINTAINABILITY FUNCTION FOR EXPONENTIAL DISTRIBUTION

Exponential distribution is simple and straightforward to handle and is quite useful to represent repair times. Its probability density function with respect to repair times is defined by

$$f_r(t) = \mu e^{-\mu t} \qquad (7.2)$$

where $f_r(t)$ is the repair time probability density function, μ is the constant repair rate or reciprocal of the mean time to repair ($MTTR$), and t is the variable repair time.

Inserting Equation 7.2 into Equation 7.1 yields

$$\begin{aligned} M_e(t) &= \int_0^t \mu e^{-\mu t}\, dt \\ &= 1 - e^{-\mu t} \end{aligned} \qquad (7.3)$$

where $M_e(t)$ is the maintainability function for exponential distribution.

Since $\mu = 1/MTTR$, Equation 7.3 becomes

$$M_e(t) = 1 - e^{-\left(\frac{1}{MTTR}\right)t} \qquad (7.4)$$

Example 7.1

Assume that the repair times of a mechanical system are exponentially distributed with a mean value of 5 hours. Calculate the probability of completing a repair in 6 hours.

Using the specified data values in Equation 7.4 yields

$$\begin{aligned} M_e(6) &= 1 - e^{-\left(\frac{6}{5}\right)} \\ &= 0.6988 \end{aligned}$$

This means there is a likelihood of approximately 70% that the repair will be completed within 6 hours.

7.4.2 MAINTAINABILITY FUNCTION FOR RAYLEIGH DISTRIBUTION

Rayleigh distribution is often used in reliability studies and it can also be used to represent corrective maintenance times. Its probability density function with respect to corrective maintenance times (i.e., repair times) is defined by

$$f_r(t) = \frac{2}{\alpha^2} t\, e^{-\left(\frac{t}{\alpha}\right)^2} \tag{7.5}$$

where $f_r(t)$ is the repair time probability density function, t is the variable repair time, and α is the distribution scale parameter.

By substituting Equation 7.5 into Equation 7.1, we get

$$M_r(t) = \int_0^t \frac{2}{\alpha^2} t\, e^{-\left(\frac{t}{\alpha}\right)^2} dt$$
$$= 1 - e^{-\left(\frac{t}{\alpha}\right)^2} \tag{7.6}$$

where $M_r(t)$ is the maintainability function for Rayleigh distribution.

7.4.3 MAINTAINABILITY FUNCTION FOR WEIBULL DISTRIBUTION

Sometimes Weibull distribution is used to represent corrective maintenance times, particularly for electronic equipment. Its probability density function with respect to corrective maintenance times is expressed by

$$f_r(t) = \frac{2}{\alpha^\theta} t^{\theta-1} e^{-\left(\frac{t}{\alpha}\right)^\theta} \tag{7.7}$$

where $f_r(t)$ is the corrective maintenance or repair time probability density function, t is the variable repair time, θ is the distribution shape parameter, and α is the distribution scale parameter.

Substituting Equation 7.7 into Equation 7.1 yields

$$M_w(t) = \int_0^t \frac{\theta}{\alpha^\theta} t^{\theta-1} e^{-\left(\frac{t}{\alpha}\right)^\theta} dt$$
$$= 1 - e^{-\left(\frac{t}{\alpha}\right)^\theta} \tag{7.8}$$

where $M_w(t)$ is the maintainability function for Weibull distribution.

At $\theta = 1$ and $\theta = 2$, Equation 7.8 reduces to Equation 7.3, for $\mu = 1/\alpha$, and (7.6), respectively.

7.4.4 MAINTAINABILITY FUNCTION FOR GAMMA DISTRIBUTION

Gamma distribution is one of the most flexible distributions and it can be used to represent various types of maintenance time data. Its probability density function with respect to repair times is defined by

$$f_r(t) = \frac{c^m}{\Gamma(m)}\, t^{m-1}\, e^{-ct} \tag{7.9}$$

where $f_r(t)$ is the repair time probability density function, t is the variable repair time, c is the distribution scale parameter, and m is the distribution shape parameter.

The gamma function, $\Gamma(m)$, is given by [3]

$$\Gamma(m) = \int_0^\infty y^{m-1}\, e^{-y}\, dy \tag{7.10}$$

By substituting Equation 7.9 into Equation 7.1, we get

$$M_g(t) = \frac{c^m}{\Gamma(m)} \int_0^t t^{m-1}\, e^{-ct}\, dt \tag{7.11}$$

where $M_g(t)$ is the maintainability function for gamma distribution.

For $m = 1$, Equation 7.11 becomes the maintainability function for the exponential distribution. In order to find, $M_g(t)$, by using the tables of the incomplete gamma function, we rewrite Equation 7.11 to the following form [11]:

$$M_g(t) = c^m\, I(t) \tag{7.12}$$

where

$$I(t) = \frac{1}{\Gamma(m)} \int_0^t t^{m-1}\, e^{-ct}\, dt \tag{7.13}$$

The mean of the gamma distributed repair times is given by

$$\beta = \frac{m}{c} \tag{7.14}$$

where β is the mean value of the gamma distributed repair times.

The standard deviation, σ, of the gamma distributed repair times is

$$\sigma = \frac{\sqrt{m}}{c} = \left(\frac{\beta}{c}\right)^{1/2} \tag{7.15}$$

7.4.5 Maintainability Function for Erlangian Distribution

For positive integer values of m, the shape parameter, the gamma distribution becomes the Erlangian distribution. In this case, Equation 7.10 yields

$$\Gamma(m) = (m-1)! \tag{7.16}$$

Thus, the probability density function of the Erlangian distribution from Equation 7.9 is

$$f_r(t) = \frac{c}{(m-1)!}(ct)^{m-1}e^{-ct} \tag{7.17}$$

where $f_r(t)$ is the repair time probability density function of the Erlangian distribution, m is the distribution shape parameter, c is the distribution scale parameter, and t is the variable repair time.

Inserting Equation 7.17 into Equation 7.1, we get

$$M_{er}(t) = 1 - \sum_{i=0}^{m-1}\left\{e^{-ct}(ct)^i / i!\right\}$$
$$= \sum_{i=m}^{\infty}\left\{e^{-ct}(ct)^i / i!\right\} \tag{7.18}$$

where $M_{er}(t)$ is the maintainability function for Erlangian distribution.

7.4.6 Maintainability Function for Normal Distribution

Normal distribution is one of the most well-known probability distributions and it can also be used to represent failed equipment repair times. Its probability density function with respect to repair times is expressed by

$$f_r(t) = \frac{1}{\sigma\sqrt{2\Pi}} \exp\left[-\frac{1}{2}\left(\frac{t-\mu}{\sigma}\right)^2\right] \tag{7.19}$$

where $f_r(t)$ is the repair time probability density function, t is the variable repair time, μ is the mean of repair times, and σ is the standard deviation of the variable repair time t around the mean value of μ.
Substituting Equation 7.19 into Equation 7.1 yields

$$M_n(t) = \frac{1}{\sigma\sqrt{2\Pi}} \int_0^t \exp\left[-\frac{1}{2}\left(\frac{t-\mu}{\sigma}\right)^2\right] dt \qquad (7.20)$$

where $M_n(t)$ is the maintainability function for normal distribution.
The mean of repair times is given by

$$\mu = \sum_{i=1}^n t_i/n \qquad (7.21)$$

where n is the number of repair times and t_i is the repair time i for $i = 1, 2, 3, \ldots, n$.
The standard deviation is given by

$$\sigma = \left[\sum_{i=1}^n (t_i - \mu)^2/(n-1)\right]^{1/2} \qquad (7.22)$$

7.4.7 Maintainability Function for Lognormal Distribution

Lognormal distribution is probably the most widely used probability distribution in maintainability work. Its probability density function with respect to repair times is defined by

$$f_r(t) = \frac{1}{(t-\gamma)\sigma\sqrt{2\Pi}} \exp\left[-\frac{1}{2}\left\{\frac{\ln(t-\gamma)-\beta}{\sigma}\right\}^2\right] \qquad (7.23)$$

where $f_r(t)$ is the probability density function of the repair times, t is the variable repair time, γ is a constant denoting the shortest time below which no repair action can be carried out, β is the mean of the natural logarithms of the repair times, and σ is the standard deviation with which the natural logarithm of the repair times are spread around the mean β.
By substituting Equation 7.23 into Equation 7.1, we get

$$M_l(t) = \int_0^t \frac{1}{(t-\gamma)\sigma\sqrt{2\Pi}} \exp\left[-\frac{1}{2}\left\{\frac{\ln(t-\gamma)-\beta}{\sigma}\right\}^2\right] dt \qquad (7.24)$$

where $M_l(t)$ is the maintainability function for lognormal distribution.
  The following relationship defines the mean:

$$\beta = \left[\ln t_1 + \ln t_2 + \ln t_3 + \cdots\cdots + \ln t_k \right]/k \qquad (7.25)$$

where k is the total number of repair times and t_i is the repair time i for $i = 1, 2, 3, \ldots., k$.
  The standard deviation, σ, is expressed by

$$\sigma = \left[\sum_{i=1}^{k} \left(\ln t_i - \beta \right)^2 / (k-1) \right]^{1/2} \qquad (7.26)$$

7.5 PROBLEMS

1. Discuss the need for maintainability.
2. Compare engineering maintainability with engineering maintenance.
3. Compare maintainability engineering with reliability engineering.
4. Define maintainability function.
5. Write down the maintainability function for an exponential distribution.
6. Assume that the repair times of an engineering system are exponentially distributed with a mean value of 6 hours. Calculate the probability of accomplishing a repair in 8 hours.
7. Prove that the maintainability function for Weibull distribution is given by Equation 7.8.
8. Prove that the mean of the gamma distributed repair times is given by Equation 7.14.
9. Prove that the maintainability function for Erlangian distribution is given by Equation 7.18.

REFERENCES

1. Latino, C.J., *Hidden Treasure: Eliminating Chronic Failures Can Cut Maintenance Cost up to 60%*, Report, Reliability Center, Hopewell, VA, 1999.
2. Kumar, U.D., New trends in aircraft reliability and maintenance measures, *Journal of Quality in Maintenance Engineering*, 5(4), 287–295, 1999.
3. *Engineering Design Handbook: Maintainability Engineering Theory and Practice*, AMCP 706-133, Department of Defense, Washington, DC, 1976.
4. Downs, W.R., Maintainability analysis versus maintenance analysis, *Proceedings of the Annual Reliability and Maintainability Symposium*, 1976, pp. 476–481.
5. Smith, D.J. and Babb, A.H., *Maintainability Engineering*, John Wiley & Sons, New York, 1973.
6. Dhillon, B.S., *Engineering Maintainability*, Gulf Publishing, Houston, TX, 1999.
7. *Engineering Design Handbook: Maintainability Guide for Design*, AMCP-706-134, Department of Defense, Washington, DC, 1972.

8. Blanchard, B.S., Verma, D., and Peterson, E.L., *Maintainability*, John Wiley & Sons, New York, 1995.

9. Dhillon, B.S., *Reliability Engineering in Systems Design and Operation*, Van Nostrand Reinhold, New York, 1983.

10. Von Alven, W.H., Ed., *Reliability Engineering*, Prentice Hall, Englewood Cliffs, NJ, 1964.

11. Pearson, K., *Tables of the Incomplete Gamma Function*, Cambridge University Press, Cambridge, UK, 1934.

8 Maintainability Tools and Specific Maintainability Design Considerations

8.1 INTRODUCTION

Many methods and techniques have been developed to perform various types of reliability and quality analyses. Some of these approaches have been successfully used in the maintainability area as well. These approaches include fault tree analysis; cause and effect diagram; failure modes, effects, and criticality analysis (FMECA); and total quality management.

An effective engineering design (i.e., cost-effective and supportable design) takes into account the maintainability considerations that arise during the equipment or item life cycle phases. This requires careful planning and a systematic effort to bring attention to maintainability design factors such as maintainability allocation, maintainability evaluation, and maintainability design characteristics. Many of these factors involve subfactors such as interchangeability, standardization, modularization, accessibility, testing and checkout, human factors, and safety. In every aspect of maintainability design interchangeability, standardization, modularization, and accessibility are important considerations [1,2].

This chapter presents a number of methods for performing various types of maintainability analysis and various aspects of specific maintainability design considerations.

8.2 FAULT TREE ANALYSIS

This method was developed at the Bell Laboratories in the early 1960s to evaluate the reliability and safety of the Minuteman Launch Control System [3]. Fault tree analysis (FTA) starts by defining the undesirable state (event) of the system or item under consideration and then analyzes the system to determine all possible situations that can result in the occurrence of the undesirable event. Thus, it identifies all possible failure causes at all possible levels associated with a system as well as the relationship between causes. FTA can be used to analyze various types of maintainability-related problems.

FTA uses various types of symbols [3]. Four commonly used symbols in fault tree construction are shown in Figure 8.1. The circle denotes a basic fault event or the failure of an elementary component. The event's occurrence probability and failure and repair rates are normally obtained from empirical data. The rectangle denotes a fault event that results from the combination of fault events through the input of a logic gate.

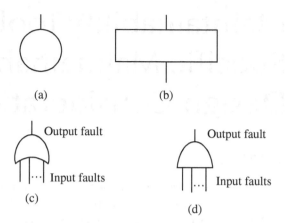

(a)   (b)

(c)

(d)

FIGURE 8.1 Four commonly used fault tree symbols: (a) circle, (b) rectangle, (c) OR gate, and (d) AND gate.

The OR gate denotes that an output fault occurs if one or more of the input fault events occur. Finally, the AND gate denotes that an output fault event occurs only if all the input fault events occur. The probabilities of the occurrence of the output fault events of logic gates (OR and AND) are given by

OR gate

$$P(E_0) = 1 - \prod_{i=1}^{n} \left\{ 1 - P(E_i) \right\} \tag{8.1}$$

where $P(E_0)$ is the probability of occurrence of the OR gate output fault event, E_0, n is the number of independent input fault events, and $P(E_i)$ is the probability of occurrence of input fault event E_i for $i = 1, 2, 3, \ldots, n$.

AND gate

$$P(Y_0) = \prod_{i=1}^{n} P(Y_i) \tag{8.2}$$

where
$P(Y_0)$ is the probability of occurrence of the AND gate output fault event, Y_0 and $P(Y_i)$ is the probability of occurrence of input fault event Y_i for $i = 1, 2, 3, \ldots, n$.

Needless to say, FTA can be used to analyze various maintainability-related problems. The following example demonstrates its application to a maintainability-related problem:

Example 8.1

A workshop repairs failed engineering equipment and it will not be able to repair a given piece of equipment because of the factors listed below.

   A: Skilled manpower is unavailable.
   B: Equipment is too damaged to repair.

C: Repair facilities or tools are unavailable.
D: There are no spare parts.

Furthermore, either of the following two factors can result in the unavailability of spare parts:

E: Parts are no longer available in the market.
F: Parts are out of stock.

In addition, the unavailability of skilled manpower can be caused by either of the following two factors:

G: Poor planning.
H: Labor shortage.

Develop a fault tree for this undesired event: the equipment will not be repaired by a given point in time. Calculate the probability of the occurrence of the undesired event if the probabilities of occurrence of factors B, C, E, F, G, and H are 0.03, 0.02, 0.04, 0.05, 0.06, and 0.07, respectively.

For this example, the fault tree shown in Figure 8.2 was developed using the symbols from Figure 8.1 . Single capital letters in the figure diagram donate corresponding fault

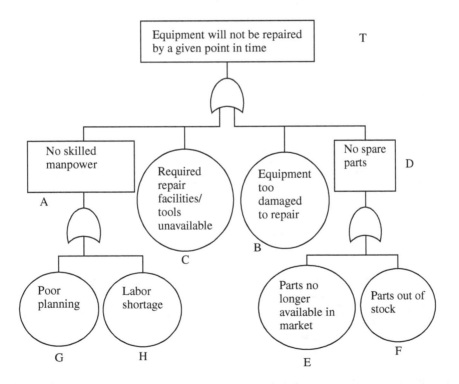

FIGURE 8.2 Fault tree for Example 8.1.

events. By substituting the given data values into Equation 5.1, the probability of occurrence of event A (i.e., skilled manpower is unavailable) is given by

$$P(A) = 1 - (1 - 0.06)(1 - 0.07)$$
$$= 0.1258$$

Similarly, by substituting the specified data values into Equation 5.1, the probability of occurrence of fault event D (i.e., no spare parts) is given by

$$P(D) = 1 - (1 - 0.04)(1 - 0.05)$$
$$= 0.088$$

With the above two calculated values and the data in Equation 5.1, the probability of occurrence of the undesired event (i.e., equipment will not be repaired by a given point in time) is given by

$$P(T) = 1 - (1 - 0.088)(1 - 0.02)(1 - 0.03)(1 - 0.1258)$$
$$= 0.2421$$

where $P(T)$ is the probability that the equipment will not be repaired by a given point in time.

Figure 8.3 presents the Figure 8.2 fault tree with given and calculated fault event occurrence probability values. As per Figure 8.3, the probability that equipment will not be repaired by a given point in time is 0.2421.

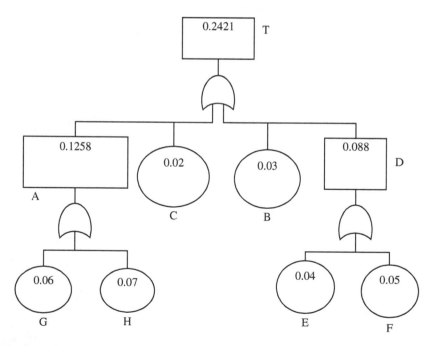

FIGURE 8.3 A fault tree for Example 8.1 with given and calculated fault event occurrence probability values.

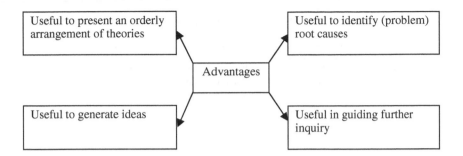

FIGURE 8.4 Important advantages of the cause and effect diagram.

8.3 CAUSE AND EFFECT DIAGRAM

This is a deductive analysis approach that can be quite useful in maintainability work. In the published literature, this method is also known as a fishbone diagram because it resembles the skeleton of a fish, or as an Ishikawa diagram, after its originator, K. Ishikawa of Japan [4]. A cause and effect diagram uses a graphic fishbone for depicting the cause and effect relationships between an undesired event and its associated contributing causes.

The right side (i.e., the fish head or the box) of the diagram represents the effect (i.e., the problem or the undesired event), and left of this, all possible causes of the problem are connected to the central fish spine. The basic steps involved in developing a cause and effect diagram are:

- Establish a problem statement or highlight the effect to be investigated.
- Brainstorm to identify all possible causes for the problem under study.
- Group major causes into categories and stratify them.
- Construct the diagram by linking the causes under appropriate process steps and write down the effect or problem in the diagram box (i.e., the fish head) on the right side.
- Refine cause categories by asking questions such as "What causes this?" and "What is the reason for the existence of this condition?"

Some of the important advantages of the cause and effect diagram are shown in Figure 8.4.

A well-developed cause and effect diagram can be an effective tool to identify possible maintainability-related problems [2].

8.4 FAILURE MODES, EFFECTS, AND CRITICALITY ANALYSIS (FMECA)

This method grew out of failure mode and effects analysis (FMEA), discussed in Chapter 4. When FMEA evaluates the failure criticality (i.e., the failure effect severity and its occurrence probability), the method is called FMECA and the failure modes are assigned priorities [5].

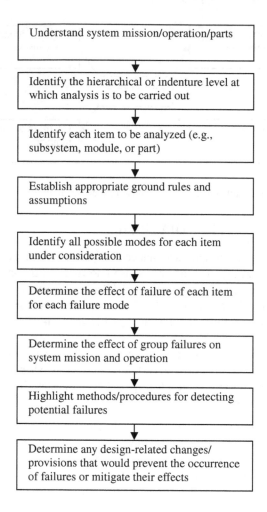

FIGURE 8.5 Basic steps used for performing FMECA.

The basic steps used to perform FMECA are shown in Figure 8.5 [5]. Some of the useful information concerning FMECA and corresponding sources for obtaining it are presented in Table 8.1 [2, 5].

Some of the advantages of FMECA are as follows [6]:

- Easy to understand
- A useful tool for identifying all possible failure modes and their effects on the mission, the system, and personnel
- Useful for making design comparisons
- A visibility tool for managers and others
- Useful for generating data for application in system safety and maintainability analyses
- A systematic method for classifying hardware failures

TABLE 8.1
Useful Information Concerning FMECA and Sources for Obtaining It

No.	Information	Source
1	• Item identification numbers	• Parts list for the product or system
2	• Product and system function	• Customer requirements or the design engineer
3	• Mission phase and operational mode	• Design engineer
4	• Item nomenclature and functional specifications	• Design engineer or parts list
5	• Item failure modes, causes, and rates	• Factory database or field experience database
6	• Provisions and design changes to prevent or compensate for failures	• Design engineer
7	• Failure detection method(s)	• Design engineer or maintainability engineer
8	• Failure probability and severity classification	• Safety engineer
9	• Failure effects	• Design engineer, safety engineer, or reliability engineer

- Useful to analyze small, large, and complex systems effectively
- Starts from the level of greatest detail and works in the upward direction
- Useful for generating input data for application in test planning
- An effective tool for improving communication among design interface personnel

8.5 TOTAL QUALITY MANAGEMENT

Total quality management (TQM) is a philosophy of pursuing continuous improvement in every process through the integrated or team efforts of all people in an organization. It has proven to be quite useful to organizations in pursuit of improving the maintainability of their products. The term *total quality management* was coined by Nancy Warren, an American behavioral scientist, in 1985 [7].

Two fundamental principles of TQM are continuous improvement and customer satisfaction, and its seven important elements are listed below [8]:

- Management commitment and leadership
- Team effort
- Supplier participation
- Cost of quality
- Training
- Statistical tools
- Customer service

TQM can be implemented by following the five basic steps shown in Figure 8.6 [2].

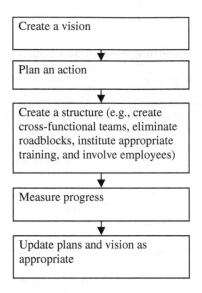

FIGURE 8.6 Basic steps for implementing TQM.

Many organizations have experienced various difficulties in implementing TQM. Some of those difficulties are failure of top management to devote adequate time to the effort, failure of senior management to delegate decision-making authority to lower organizational levels, insufficient allocation of resources for training and developing manpower, and management insisting on implementing processes in a way employees find unacceptable [9].

8.6 MAINTAINABILITY DESIGN FACTORS

Goals of maintainability design include reducing support costs, increasing ease of maintenance, minimizing preventive and corrective maintenance tasks, and minimizing the logistical burden through resources (e.g., spare parts, repair staff, and support equipment) required for maintenance and support [1].

Maintainability design factors that are most frequently addressed are presented in Table 8.2 [1]. Additional factors are listed below [1]:

- Standardization
- Modular design
- Interchangeability
- Ease of removal and replacement
- Lubrication
- Servicing equipment
- Skill requirements
- Indication and location of failures
- Work environment
- Required number of personnel

TABLE 8.2
Most Frequently Addressed Maintainability Design Factors
Ranked in Descending Order

No.	Maintainability Design Factor
1	• Accessibility
2	• Test points
3	• Controls
4	• Labeling and coding
5	• Displays
6	• Manuals, checklists, charts, and aids
7	• Test equipment
8	• Tools
9	• Connectors
10	• Cases, covers, and doors
11	• Mounting and fasteners
12	• Handles
13	• Safety factors

- Adjustments and calibrations
- Functional packaging
- Weight
- Cabling and wiring
- Fuses and circuit breakers
- Installation
- Illumination
- Training requirements
- Test adopters and test hook-ups

8.7 STANDARDIZATION AND MODULARIZATION

Standardization is the attainment of maximum practical uniformity in product design [10,11]. More specifically, it restricts to a minimum the variety of components that a product will require. There are many goals of standardization. Some of the important ones are shown in Figure 8.7 [2].

Standardization should be the main goal of design, because the use of nonstandard components may result in increased maintenance and lower reliability. Nonetheless, past experiences indicate that the lack of standardization is usually due to poor communication among design engineers, users, contractors, subcontractors, and so on [12]. Some of the advantages of standardization are [2]:

- Reduction in the danger of using the wrong parts
- Elimination in the need for special or close-tolerance parts
- Reduction in wiring and installation errors because of variations in characteristics of similar items
- Better reliability

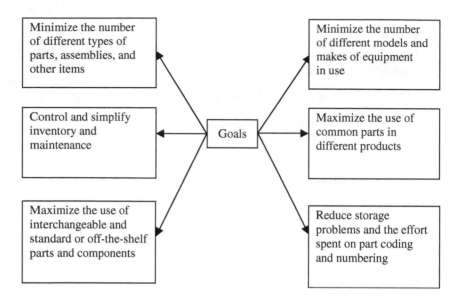

FIGURE 8.7 Important standardization goals.

- Reduction in the occurrence of accidents caused by wrong or unclear procedures
- Reduction of manufacturing costs, design time, and maintenance cost and time
- Reduction in procurement, stocking, and training problems

Modularization is the division of a product into functionally and physically distinct units to allow easy removal and replacement. The degree of modularization of a product is dictated by factors such as cost, practicality, and function. Some useful guidelines associated with designing modularized products are [11,12]:

- Divide the equipment or item under consideration into many modular parts or units.
- Aim to make modules and parts as uniform as possible with respect to size and shape.
- Aim to make each module capable of being inspected independently.
- Design the equipment so that a single person can replace any failed part without any difficulty.
- Aim to design modules for maximum ease of operational testing when they are removed from the actual equipment or system.
- Make each modular unit light and small so that a person can carry and handle it in an effective manner without any problem.
- Follow an integrated approach to design. More specifically, consider design, modularization, and material problems simultaneously.
- Emphasize modularization for forward levels of maintenance as much as possible to increase operational capability.
- Design control levers and linkages to allow easy disconnection from components. This will make it easier to replace components.

Some important advantages of modularization are ease of dividing up maintenance responsibilities; relative ease of maintaining a divisible configuration; simpler new equipment design; shorter design time; less costly and less time consuming training of maintenance personnel; lower levels of skill needed to replace modular units in the field as well as the need for fewer tools; reduction in equipment downtime as recognition, isolation, and replacement of faulty items become easier; and ease of modifying existing equipment with the latest functional units by simply replacing their older equivalents [11].

8.8 SIMPLIFICATION AND ACCESSIBILITY

Simplification is the most important element of maintainability and it is also probably the most difficult to achieve. Simplification should be the constant objective of design, and a good design engineer includes pertinent functions of a system or product into the design itself and makes use of as few components as good design practices will permit.

Accessibility is the relative ease with which an item can be reached for repair, replacement, or service. Poor accessibility is a frequent cause of ineffective maintenance. For example, according to a U.S. Army document, gaining access to equipment is probably second only to fault isolation as a time-consuming maintenance task [1].

Some of the factors that affect accessibility are the visual needs of personnel performing the tasks, the item's location and environment, the distance to be reached to access the item, the types of maintenance tasks to be performed through the access opening, the frequency with which the access opening is entered, the danger associated with use of the access opening, specified time requirements for performing the tasks, the clothing worn by the maintenance personnel, the types of tools and accessories required to perform the tasks, the mounting of items behind the access opening, and work clearances necessary for carrying out the tasks [1,11].

It should be added that an item being readily accessible does not in itself guarantee overall cost-effectiveness and ease of maintenance under consideration.

8.9 INTERCHANGEABILITY AND IDENTIFICATION

Interchangeability means including as an intentional aspect of design that any item can be replaced within a product by any similar item. Interchangeability is made possible through standardization and is an important maintainability design factor. Three basic principles of interchangeability are:

- In items and products requiring frequent servicing and replacement of parts, each part must be interchangeable with another similar part.
- Liberal tolerances must exist.
- Strict interchangeability could be uneconomical in items and products that are expected to operate without any part replacement.

In order to achieve maximum interchangeability of parts, units, and items, a design engineer must ensure the following in a system under consideration [1]:

- Existence of functional interchangeability when physical interchangeability is a design characteristic
- Availability of sufficient information in task instructions and number plate identification for allowing users to decide with confidence whether two similar parts are interchangeable
- No change in methods of connecting and mounting when there are part or unit modifications
- Avoidance of differences in size, mounting, shape, etc.
- Availability of adapters for making physical interchangeability possible when total interchangeability is not practicable
- Total interchangeability of identical parts, identified as interchangeable through some appropriate identification system.

However, when functional interchangeability is not required, there is no need to have physical interchangeability.

Identification is concerned with labeling or marking of parts, controls, and test points to facilitate tasks such as repair and replacement. When parts, controls, and test points are not identified effectively, the performance of maintenance tasks becomes more difficult, takes longer, and increases the chances for making errors. Types of identification include equipment identification and part identification. Additional information on identification is available in References 1 and 2.

8.10 PROBLEMS

1. Write an essay on fault tree analysis.
2. What is the difference between failure mode and effects analysis (FMEA) and failure mode, effect, and criticality analysis (FMECA)?
3. What are the main steps used for performing FMECA?
4. Discuss the following methods:
   - Cause and effect diagram
   - Total quality management
5. What are the important sources for obtaining information when performing FMECA?
6. What are the important benefits of FMECA?
7. What are the most frequently addressed maintainability design factors?
8. What are the important goals of standardization?
9. Discuss at least six important guidelines associated with designing modularized products.
10. Describe the following:
    - Interchangeability
    - Accessibility
    - Identification

REFERENCES

1. *Engineering Design Handbook: Maintainability Engineering Theory and Practice*, AMCP 706-133, Department of Defense, Washington, DC, 1976.
2. Dhillon, B.S., *Engineering Maintainability*, Gulf Publishing, Houston, TX, 1999.
3. Dhillon, B.S. and Singh, C., *Engineering Reliability: New Techniques and Applications*, John Wiley & Sons, New York, 1981.
4. Ishikawa, K., *Guide to Quality Control*, Asian Productivity Organization, Tokyo, 1976.
5. Bowles, J.B. and Bonnell, R.D., Failure, mode, effects and criticality analysis, in *Tutorial Notes: Annual Reliability and Maintainability Symposium, 1994*. Evans Associates, Durham, NC, 1994, pp. 1–34.
6. Dhillon, B.S., *Systems Reliability, Maintainability, and Management*, Petrocelli Books, New York, 1983.
7. Walton, M., Deming *Management at Work*, Putnam, New York, 1990.
8. Burati, J.L., Matthews, M.F., and Kalidindi, S.N., Quality management organizations and techniques, *Journal of Construction Engineering and Management*, 118, 112–128, 1992.
9. Gevirtz, C.D., *Developing New Products with TQM*, McGraw-Hill, New York, 1994.
10. Ankenbrandt, F.L., et al., *Maintainability Design*, Engineering Publishers, Elizabeth, NJ, 1963.
11. *Engineering Design Handbook: Maintainability Guide for Design*, AMCP-706-134, Department of Defense, Washington, DC, 1972.
12. Rigby, L.V., et al., *Guide to Integrated System Design for Maintainability*, Report No. ASD-TR-61-424, U.S. Air Force Systems Command, Wright-Patterson Air Force Base, OH, 1961.

REFERENCES

9 Maintainability Management and Costing

9.1 INTRODUCTION

Just like in any other area of engineering, management plays an important role in the practice of maintainability engineering. Its tasks range from simply managing maintainability personnel to effective execution of technical maintainability tasks. Maintainability management can be examined from different perspectives such as management of maintainability as an engineering discipline, the place of the maintainability function within the organizational structure, and the role maintainability plays at each phase in the life cycle of system and product under development [1].

Maintainability costing can be examined from different perspectives including the cost of performing the maintainability function and the cost of maintaining a product in the field. In regard to the latter, equipment's operation and support cost can account for as much as 75% of its total life cycle cost [2]. Obviously, this cost must be reduced to a minimal level to make the equipment cost-effective. Nonetheless, experiences indicate that 60 to 70% of the projected life cycle cost of a product can sometimes be locked in by the completion of the preliminary design phase.

9.2 MAINTAINABILITY MANAGEMENT TASKS DURING THE PRODUCT LIFE CYCLE

During the product life cycle, as maintainability issues arise, various types of maintainability management-related tasks are performed. An effective maintainability program incorporates a dialogue between the manufacturer and user throughout the product life cycle, which can be divided into four distinct phases as shown in Figure 9.1 [3].

The concept development phase is the first phase of the product life cycle. During this phase the product operational needs are translated into a set of operational requirements and high-risk areas are highlighted. The main maintainability management task during this phase is concerned with determining the product effectiveness requirements as well as determining, from the product's purpose and intended operation, the required field support policies and other provisions.

The validation phase is the second phase of the product life cycle. Some of the maintainability management tasks associated with this phase are developing a maintainability program plan that satisfies contractual requirements;

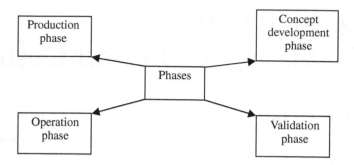

FIGURE 9.1 Product life cycle phases.

establishing maintainability incentives and penalties; performing maintainability allocations and predictions; developing a plan for maintainability testing
and demonstration; coordinating and monitoring maintainability efforts
throughout the organization; participating in design reviews; establishing maintainability policies and procedures for the validation phase and the subsequent
full-scale engineering effort; developing a planning document for data collection,
analysis, and evaluation; and providing appropriate assistance to maintenance
engineering in areas such as performing maintenance analysis and developing
logistic policies [3].

The production phase is the third phase of the product life cycle. Some of the
maintainability management tasks associated with this phase are [4]:

- Evaluating all proposals for changes in regard to their impact on
  maintainability
- Evaluating production test trends from the standpoint of adverse effects
  on maintainability requirements
- Participating in the development of appropriate controls for errors, process
  variations, and other problems that may affect maintainability directly or
  indirectly
- Monitoring production processes
- Ensuring the eradication of all shortcomings that may degrade maintainability

The operation phase is the final phase of the product life cycle. No specific
maintainability management-related tasks are involved with this phase, but the phase
is probably the most significant because during this period the product's true logistic
support and cost-effectiveness are demonstrated. Thus, essential maintainability-
related data can be collected for use in future applications [4].

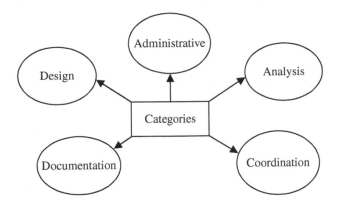

FIGURE 9.2 Maintainability organization function categories.

9.3 MAINTAINABILITY ORGANIZATION FUNCTIONS

A maintainability organization performs a wide variety of functions that can be grouped under five distinct categories as shown in Figure 9.2 [1,4].

The administrative category includes functions such as [1,5]:

- Preparing budgets and schedules
- Assigning maintainability-related responsibilities
- Monitoring the maintainability organization's output
- Providing maintainability training
- Developing and issuing policies and procedures for application in maintainability efforts
- Taking part in program management and design reviews
- Organizing the maintainability effort
- Acting as a liaison with higher-level management and other concerned bodies

Some of the important design category functions are presented in Table 9.1 [1,4]. The analysis category includes functions such as [1,5]:

TABLE 9.1
Important Functions in the Design Category

No.	Function
1	Participating in the development of maintainability design criteria and guidelines
2	Reviewing product design with respect to maintainability
3	Approving design drawings from the maintainability standpoint
4	Preparing maintainability-related design documents
5	Providing consulting services to professionals such as design engineers

---

- Reviewing product and system specification documents from the standpoint of maintainability requirements
- Performing analysis of maintainability data obtained from the field and other sources
- Conducting maintainability allocation and prediction studies
- Developing maintainability demonstration documents
- Taking part in or conducting required maintenance analysis
- Participating in product engineering analysis to safeguard maintainability interests

The coordination category includes functions such as interfacing with product engineering and other engineering disciplines; coordinating maintainability training activities for all people involved; coordinating with professional societies, governments, and trade associations on maintainability-related matters; and acting as a liaison with subcontractors on maintainability-related issues [1,4].

Some of the functions in the documentation category are [1,4,5]:

- Documenting the results of maintainability analysis and trade-off studies
- Developing maintainability-related data and feedback reports
- Documenting information concerning maintainability management
- Documenting maintainability design review results
- Developing and maintaining a maintainability data bank
- Developing and maintaining handbook data and information on maintainability-related issues
- Establishing and maintaining a library facility that contains important maintainability documents and information

9.4 MAINTAINABILITY PROGRAM PLAN

This is an important document that contains maintainability-related information concerning a project under consideration. It is developed either by the product or system manufacturers or the user, depending on factors such as the nature of the project and the philosophy of the decision makers. Some of the important elements of a maintainability program plan are [2,4]:

Objectives: These are basically the descriptions of the overall requirements for the maintainability program and goals of the plan.

Policies and procedures: Their main purpose is to assure customers that the group implementing the maintainability program will perform its assigned task in an effective manner. Under the policies and procedures, the management's overall policy directives for maintainability are also referenced or incorporated. The directives address items such as data collection and analysis, maintainability demonstration methods, participation in design reviews and evaluation, and methods to be employed for maintainability allocation and prediction.

Organization: A detailed organizational breakdown of the maintainability group involved in the project is provided along with the overall structure of the enterprise. In addition, information concerning the background and experience of the maintainability group personnel, a work breakdown structure, and a list of the personnel assigned to each task are provided.

Maintainability program tasks: Each program task, task schedule, major milestones, expected task output results, projected cost, and task input requirements are described in detail.

Maintainability design criteria: This part of the plan discusses specific maintainability-related design features applicable to the item under consideration. In addition, the description may relate to qualitative and quantitative factors concerning areas such as interchangeability, accessibility, parts selection, or packaging.

Organizational interfaces: This section describes the lines of communication and the relationships between the maintainability group and the overall organization. Some of the areas of interface are product engineering, design, testing and evaluation, reliability engineering, human factors, and logistic support, as well as suppliers and customers.

Technical communications: This section briefly discusses every deliverable item and their associated due dates.

Program review, evaluation, and control: This section discusses the methods and techniques to be employed for technical design reviews, program reviews, and feedback and control. Also, it describes a risk management plan and discusses the evaluation and incorporation of proposed changes and corrective actions to be taken in given situations.

Maintenance concept: This section discusses basic maintenance requirements of the product under consideration and issues such as organizational responsibilities, qualitative and quantitative objectives for maintenance and maintainability, operational and support concepts, test and support equipment criteria, and spare and repair part factors.

Subcontractor and supplier activity: This section discusses the organization's relationships with suppliers and subcontractors connected to the maintainability program. In addition, it outlines the procedures to be employed for review and control within the framework of those relationships.

References: This section lists all documents related to the maintainability requirements (e.g., applicable standards, specifications, and plans).

9.5 MAINTAINABILITY DESIGN REVIEWS

Design reviews are a critical element of modern design practices and they are conducted during the product design phase. The primary objective of design reviews is to determine the progress of the ongoing design effort as well as to ensure the application of correct design practices. The design review team members assess potential and existing problems in various areas concerned with the product under consideration including maintainability. Many maintainability-related issues require careful attention during design reviews. Some of these issues are [4,6–8]:

- Maintainability prediction results
- Conformance to maintainability design specifications
- Maintainability trade-off study results

- Design constraints and specified interfaces
- Maintainability demonstration test data
- Selection of parts and materials
- Use of on-line repair with redundancy
- Use of automatic test equipment
- Verification of maintainability design test plans
- Identified maintainability problem areas and proposed corrective actions
- Failure mode and effect analysis results
- Physical configuration and layout drawings and schematic diagrams
- Maintainability assessments using test data
- Use of unit replacement approach
- Assessments of maintenance and supportability
- Use of built-in monitoring and fault-isolation equipment
- Corrective actions taken and proposed
- Maintainability test data obtained from experimental models and bread-boards

9.6 MAINTAINABILITY INVESTMENT COST ELEMENTS

Maintainability is an important factor in the total cost of equipment because an increase in maintainability can lead to lower operation and maintenance costs. There are many ways of increasing maintainability including incorporating discard-at-failure maintenance, increasing self-checking features, increasing the use of automatic test equipment, designing in built-in test points, providing easy access for maintenance, using reduced-maintenance parts, and improving troubleshooting manuals [1]. Nonetheless, many elements of investment cost are related to maintainability. Some of these elements are the costs of repair parts, prime equipment, training, data, system engineering management, new operational facilities, system test and evaluation, and support equipment [1].

9.7 LIFE CYCLE COSTING

The life cycle cost is the sum of all costs incurred during the life span of an item. The term *life cycle costing* first appeared in 1965 in a document entitled "Life Cycle Costing in Equipment Procurement" prepared by the Logistics Management Institute for the United States Department of Defense [9]. Maintainability is an important factor in an item's life cycle cost. More specifically, a product's operation and maintenance costs are a major element of its life cycle cost.

Major steps involved in life cycle costing with respect to product procurement are shown in Figure 9.3 [10]. Some of the major benefits of life cycle costing are as follows [4]:

- It is a useful tool for comparing the cost of competing projects and products.
- It is a useful tool for making decisions associated with equipment replacement.

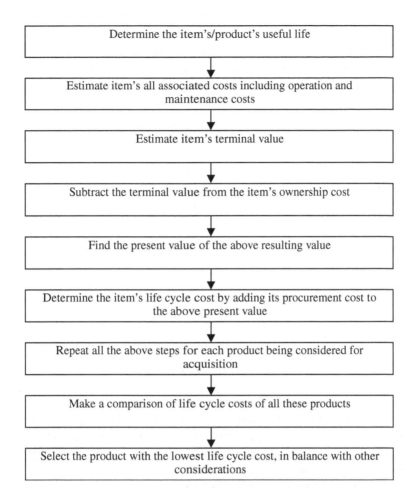

FIGURE 9.3 Life cycle costing steps with respect to product procurement.

- It is a useful tool for controlling program costs.
- It is a useful tool for selecting among competing contractors.
- It is a useful tool for conducting planning and budgeting.

Some of the disadvantages of life cycle costing are that it is time consuming, expensive, and it is a trying task to collect required data [4].

9.8 LIFE CYCLE COST ESTIMATION MODELS

Many life cycle cost models have been developed to estimate the life cycle or total cost of a product [11]. They vary in the methods they employ to determine many of the major costs used in the calculation. This section presents three mathematical models to estimate the life cycle cost of an item.

9.8.1 Life Cycle Cost Estimation Model 1

In this model, the life cycle cost (LCC) of a product is assumed to be made up of four major components [12]: conceptual phase cost (CPC), definition phase cost (DPC), acquisition phase cost (APC), and operational phase cost (OPC). Thus,

$$LCC_p = CPC + DPC + APC + OPC \qquad (9.1)$$

where LCC_p is the product, item, or system life cycle cost.

The CPC and DPC are relatively small in comparison to the costs of the acquisition and operational phases. These costs are essentially labor effort costs.

The APC is expressed by

$$APC = CPM + CPA + CPS + CSE \qquad (9.2)$$

where CPM is the cost of program management, *CPA* is the cost of personal acquisition, *CPS* is the cost of the prime system, and *CSE* is the cost of support equipment.

The OPC is defined by

$$OPC = OAE + FOE + MC \qquad (9.3)$$

where OAE is the operational administrative expense, FOE is the functional operating expense, and MC is the maintenance cost. The OAE is made up of spares inventory, investment and holding, and administrative and operational program management costs. The two main components of the functional operating expense are operational manning and consumables costs.

The MC is composed of the following elements:

- Cost of repairs and spare parts
- Cost of maintenance facilities
- Cost of maintenance personnel
- Cost of maintenance consumables
- Cost of personnel replacement
- Cost of equipment downtime

9.8.2 Life Cycle Cost Estimation Model 2

This model breaks down system or item life cycle cost into two main components: the recurring cost associated with the system (REC) and the nonrecurring cost associated with the system (NREC). Thus, the system life cycle cost is given by [13]

$$LCC_S = REC + NREC \qquad (9.4)$$

where LCC_S is the system life cycle cost.

The NREC is expressed by

$$NREC = \sum_{i=1}^{10} NREC_i \qquad (9.5)$$

where $NREC_i$ is the ith nonrecurring cost for $i = 1$ (training cost), $i = 2$ (acquisition cost), $i = 3$ (installation cost), $i = 4$ (support cost), $i = 5$ (transportation cost), $i = 6$ (reliability and maintainability improvement cost), $i = 7$ (research and development cost), $i = 8$ (life cycle costing management cost), $i = 9$ (test equipment cost), and $i = 10$ (equipment qualification approval cost).

Similarly, the REC is given by

$$REC = \sum_{i=1}^{5} REC_i \qquad (9.6)$$

where REC_i is the ith recurring cost for $i = 1$ (operating cost), $i = 2$ (inventory cost), $i = 3$ (support cost), $i = 4$ (labor cost), and $i = 5$ (maintenance cost).

9.8.3 LIFE CYCLE COST ESTIMATION MODEL 3

This model was first used by the U.S. Army Material Command to estimate the life cycle costs of new equipment or systems [2,4,14–16]. The equipment life cycle cost is expressed by

$$LCC_e = C_{rd} + C_i + C_{om} \qquad (9.7)$$

where LCC_e is the equipment or system life cycle cost, C_{rd} is the research and development cost, C_i is the investment cost, and C_{om} is the operations and maintenance cost.

The research and development cost, C_{rd}, is given by

$$C_{rd} = \sum_{i=1}^{5} C_i \qquad (9.8)$$

where C_1 is the advanced research and development cost; C_2 is the engineering development and test cost, for example, the cost of engineering models and of testing; C_3 is the engineering data cost; C_4 is the program management cost; and C_5 is the engineering design cost. It includes the costs of reliability, maintainability, system engineering, human factors, electrical design, mechanical design, producibility, and logistic support analysis.

The components of the investment cost, C_i, are as follows:

Construction cost: This includes the costs of manufacturing facilities, operational facilities, test facilities, and maintenance facilities.

Manufacturing cost: This includes the costs of manufacturing engineering, fabrication, quality control, tools and test equipment, assembly, tests and inspections, packing and shipping, and materials.

Initial logistic support cost: This includes the costs of test and support equipment, program management, first destination transportation, technical data preparation, initial spare and repair parts, initial inventory, provisioning, and initial training and training equipment.

The operations and maintenance cost, C_{om}, is given by

$$C_{om} = \sum_{i=1}^{4} C_{omi} \tag{9.9}$$

where C_{om1} is the modification cost; C_{om2} is the disposal cost; C_{om3} is the operations cost, which includes the costs of operational facilities, operations manpower, operator training, and support and handling equipment; and C_{om4} is the maintenance cost. This includes the costs of maintenance personnel, maintenance facilities, maintenance training, spare and repair parts, transportation and handling, technical data, and maintenance of test and support equipment.

Example 9.1

An owner of a trucking transport company is considering buying a truck. Two manufacturers, X and Y, are bidding to sell the truck. Data for trucks produced by the both manufacturers are presented in Table 9.2. Determine which of the two trucks is more beneficial to buy with respect to their life cycle costs.

TABLE 9.2
Life Cycle Cost-Related Data for Trucks Produced by Manufacturers X and Y

No.	Data Description	Manufacturer X's Truck	Manufacturer Y's Truck
1	Procurement cost	$130,000	$158,000
2	Annual operating cost	$25,000	$21,000
3	Annual failure rate	0.05 failures	0.06 failures
4	Disposal cost	$2,000	$2,500
5	Expected useful life	10 years	10 years
6	Annual cost of money (i.e., interest rate)	6%	6%
7	Expected cost of a failure	$2,000	$1,500

Manufacturers X's Truck

The annual expected failure cost is

$$AFC_X = (2,000)(0.05) = \$100$$

From Reference 11, the present value of the sum of uniform payments made at the end of, for example, K years, is given by

$$PV = P\left[\frac{1-(1+i)^{-K}}{i}\right] \tag{9.10}$$

where P is the uniform amount of payment made at the end of each year and i is the interest rate per year.

By substituting the above calculated value and the data given in Table 9.2 into Equation 9.10, we get

$$PFC_X = (100)\left[\frac{1-(1+0.06)^{-10}}{0.06}\right]$$
$$= \$736.01$$

where PFC_X is the present value of manufacturer X's truck failure cost.

With the specified data values in Equation 9.10, the present value of manufacturer X's truck operating cost is

$$POC_X = (25,000)\left[\frac{1-(1+0.06)^{-10}}{0.06}\right]$$
$$= \$184,002.18$$

From Reference 11, the present values of a single payment made after K years is given by

$$PV_S = \frac{P_K}{(1+i)^K} \tag{9.11}$$

where PV_S is the present value of a payment made after K years and P_K is the payment to be made after K years.

Using the data given in Equation 9.11 give the present value of manufacturer X's truck disposal cost:

$$PDC_X = \frac{(2,000)}{(1+0.06)^{10}}$$
$$= \$1116.79$$

Adding the above three calculated costs to the procurement cost, the life cycle cost of manufacturer X's truck is

$$LCC_X = 736.01 + 184,002.18 + 1,116.79 + 130,000$$
$$= \$315,854.98$$

Manufacturers Y's Truck
The annual expected failure cost is

$$AFC_y = (1,500)(0.06)$$
$$= \$90$$

Using the above calculated value and the specified data values in Equation 9.10 yields

$$PFC_y = (90)\left[\frac{1-(1+0.06)^{-10}}{0.06}\right]$$
$$= \$662.41$$

where PFC_y is the present value of manufacturer Y's truck failure cost.
    Inserting the given data values into Equation 9.10, the present value of manufacturer Y's truck operating cost is

$$POC_y = (21,000)\left[\frac{1-(1+0.06)^{-10}}{0.06}\right]$$
$$= \$154,561.83$$

With the specified data values in Equation 9.11, the present value of manufacturer Y's truck disposal cost is

$$PDC_y = \frac{(2,500)}{(1+0.06)^{10}}$$
$$= \$1,395.99$$

Adding the above three calculated costs to the procurement cost, the life cycle cost of manufacturer Y's truck is

$$LCC_y = 662.41 + 154,561.93 + 1,395.99 + 158,000$$
$$= \$314,620.23$$

Examining manufacturer X's and Y's truck life cycle costs reveals that manufacturer Y's truck will be more beneficial to buy.

9.9 PROBLEMS

1. Define the term *maintainability management*.
2. Discuss maintainability management–related tasks during the following phases of the product life cycle:
   - Concept development
   - Validation
3. Discuss important maintainability organization functions.
4. What is a maintainability program plan?
5. Discuss at least 10 important elements of a maintainability plan.
6. List at least 12 maintainability-related issues that require careful attention during product design reviews.
7. Discuss elements of investment cost related to maintainability.
8. Discuss important life cycle costing steps with respect to product procurement.
9. What are the important advantages of the life cycle costing concept?
10. A company is considering procuring an engineering system. Two manufacturers, A and B, are bidding to sell the system under consideration. Data for systems produced by the both manufacturers are presented in Table 9.3. Determine which of the two systems is more beneficial to buy with respect to their life cycle cost.

TABLE 9.3
Life Cycle Cost-Related Data for Systems Produced by Manufacturers A and B

No.	Data Description	Manufacturer A's System	Manufacturer B's System
1	Procurement cost	$200,000	$240,000
2	Annual operating cost	$24,000	$22000
3	Annual failure rate	0.02 failures	0.04 failures
4	Disposal cost	$7,000	$8,000
5	Expected useful life	15 years	15 years
6	Annual cost of money (i.e., interest rate)	4%	4%
7	Expected cost of a failure	$3,000	$2,500

REFERENCES

1. *Engineering Design Handbook: Maintainability Engineering Theory and Practice*, AMCP-706-133, Department of Defense, Washington, DC, 1976.
2. Blanchard, B.S., Verma, D., and Peterson, E.L., *Maintainability: A Key to Effective Serviceability and Maintenance Management*, John Wiley & Sons, New York, 1995.
3. Dhillon, B.S. and Reiche, H., *Reliability and Maintainability Management*, Van Nostrand Reinhold, New York, 1985.
4. Dhillon, B.S., *Engineering Maintainability*, Gulf Publishing, Houston, TX, 1999.
5. *Maintainability Program Requirements for Systems and Equipment*, MIL-STD-470, Department of Defense, Washington, DC, 1966.
6. Patton, J.D., *Maintainability and Maintenance Management*, Instrument Society of America, Research Triangle Park, NC, 1980.
7. Pecht, M., Ed., *Product Reliability, Maintainability, and Supportability Handbook*, CRC Press, Boca Raton, FL, 1995.
8. *Engineering Design Handbook: Maintainability Guide for Design*, AMCP-706-134, Department of Defense, Washington, DC, 1972.
9. *Life Cycle Costing in Equipment Procurement*, LMI Task 4C-5, Logistic Management Institute, Washington, DC, 1965.
10. Coe, C.K., Life cycle costing by state governments, *Public Administration Review*, September/October, 564–569, 1981.
11. Dhillon, B.S., *Life Cycle Costing: Techniques, Models, and Applications*, Gordon and Breach Science Publishers, New York, 1989.
12. Strodahl, N.C. and Short, J.L., The impact and structure of life cycle costing, *Proceedings of the Annual Symposium on Reliability*, 1968, pp. 509–515.
13. Reiche, H., Life cycle cost, in *Reliability and Maintainability of Electronic Systems*, Arsenault, J.E. and Roberts, I.A., Eds., Computer Science Press, Potomac, MD, 1980, pp. 3–23.
14. *Research and Development Cost Guide for Army Material Systems*, Pamphlet No. 11-2, Department of Defense, Washington, DC, 1976.
15. *Investment Cost Guide for Army Materials Systems*, Pamphlet No. 11-3, Department of Defense, Washington, DC, 1976.
16. *Operating and Support Guide for Army Material Systems*, Pamphlet No. 11-4, Department of Defense, Washington, DC, 1976.

10 Human Factors in Maintainability

10.1 INTRODUCTION

Human factors is an important discipline of engineering and it exists because people make errors in using and maintaining machines; otherwise, it would be rather difficult to justify the discipline's existence. In the published literature the terms *human factors*, *human engineering*, *ergonomics*, and *human factors engineering* have appeared interchangeably. Human factors are a body of scientific facts concerning human characteristics (the term includes all psychosocial and biomedical considerations).

Although the modern history of human factors may be traced back to Frederick W. Taylor, who carried out various studies to determine the most suitable design of shovels, human factors have only been an important element of maintainability work since World War II [1,2]. During this war the performance of military equipment clearly proved that equipment is only as good as the individuals operating and maintaining it. This means that people play an important role in the overall success of a system. Systems may fail for various reasons including poor attention given to human factors with respect to maintainability during the design phase [3].

This chapter presents various important aspects of human factors directly or indirectly related to maintainability.

10.2 GENERAL HUMAN BEHAVIORS

Many researchers have studied human behaviors and made conclusions about many general, typical, and expected behaviors. The knowledge of such behaviors can be quite useful in maintainability work directly or indirectly. Some general human behaviors are as follows [4,5]:

- People get easily confused with unfamiliar things.
- People become complacent and less careful after successfully handling hazardous items over a lengthy period.
- People have tendency to use their hands for examining or testing.
- People usually overestimate short distances and underestimate large or horizontal distances.
- People are too impatient to take the appropriate amount of time for observing precautions.
- People expect electrical switches to move upward or to the right for turning power on.
- People have become accustomed to certain color meanings.

- People read instructions and labels incorrectly or overlook them altogether.
- In emergencies, people normally respond irrationally.
- People often estimate speed or clearance poorly.
- Peoples' attention is drawn to items such as loud noises, flashing lights, bright and vivid colors, and bright lights.
- People fail to recheck their work for errors after performing a procedure.
- People, in general, have a very little idea about their physical limitations.
- People are reluctant to admit errors or mistakes.
- People assume that an object is small enough to get hold of and is light enough to pick up.
- People are rather reluctant to admit that they do not see objects clearly, whether because of poor eyesight or inadequate illumination.
- People carry out their tasks while thinking about other things.
- People usually expect that valve handles and faucets rotate counterclockwise for increasing the flow of liquid, steam, or gas.
- People can get easily distracted by certain aspects of a product's features.
- People regard manufactured products as being safe.

10.3 HUMAN SENSORY CAPABILITIES AND BODY MEASUREMENTS

In maintainability work, there is a need for an understanding of human sensory capacities as they apply to areas such as parts identification, noise, and color coding. The five major senses possessed by humans are sight, taste, smell, touch, and hearing. Humans can sense items such as pressure, vibration, temperature, linear motion, and acceleration (shock). Three of these sensors are discussed below [2,3,6].

10.3.1 TOUCH

This complements human ability to interpret visual and auditory stimuli. In maintainability work the touch sensor may be used to relieve eyes and ears of part of the load. For example, its application could be the recognition of control knob shapes with or without using other sensors.

The use of the touch sensor in technical work is not new; it has been used for many centuries by craft workers for detecting surface irregularities and roughness. Furthermore, according to Reference 7, the detection accuracy of surface irregularities dramatically improves when the worker moves an intermediate piece of paper or thin cloth over the object surface rather than simply using his or her bare fingers.

10.3.2 SIGHT

This is another sensor that plays an important role in maintainability work. Sight is stimulated by electromagnetic radiation of certain wavelengths, often known as the visible segment of the electromagnetic spectrum. In daylight, the human eye is very sensitive to greenish-yellow light and it sees differently from different angles.

Some of the important factors concerning color with respect to the human eye are as follows:

- Normally, the eye can perceive all colors when looking straight ahead. However, with an increase in viewing angle, color perception decreases significantly.
- In poorly lit areas or at night, it may be impossible to determine the color of a small point source of light (e.g., a small warning light) at a distance. In fact, the light colors will appear to be white.
- The color reversal phenomenon may occur when one is staring, for example, at a green or red light and then glances away. In such situations, the signal to the brain may reverse the color.

Some useful guidelines for designers and others are to choose colors in such a way that color-weak people do not get confused, use red filters with a wavelength greater than 6,500 Å, and avoid placing too much reliance on color when critical tasks are to be performed by fatigued personnel [8].

10.3.3 HEARING

This sensor can also be an important factor in maintainability work, as excessive noise may lead to problems including reduction in the workers' efficiency, adverse effects on tasks, need for intense concentration or a high degree of muscular coordination, and loss in hearing if exposed for long periods. In order to reduce the effects of noise, some useful guidelines related to maintainability are as follows [3]:

- Protect maintenance personnel by issuing protective devices where noise reduction is not possible.
- Incorporate into the equipment appropriate acoustical design and mufflers and other sound-proofing devices in areas where maintenance tasks must be performed in the presence of extreme noise.
- Keep noise levels below 85 dB in areas where the presence of maintenance persons is necessary.
- Prevent unprotected repair personnel from entering areas with sound levels more than 150 dB.

10.3.4 BODY MEASUREMENTS

This information is very important in designing for maintainability since humans usually operate and maintain engineering products. It helps designers ensure that equipment and products under consideration will accommodate operating and maintenance personnel of varying weights, sizes, and shapes. In turn, these people will perform their tasks effectively.

Usually human body-related requirements are outlined in the product or system design specification, particularly when the equipment is being developed for use in

TABLE 10.1
Some Body-Related Dimensions of the U.S. Adult Population (18–79 years)

No.	Description	5th Percentile (in Inches)		95th Percentile (in Inches)	
		Female	Male	Female	Male
1	Weight	104 (lb)	126 (lb)	199 (lb)	217 (lb)
2	Seated eye height	27.4	28.4	31.0	33.5
3	Standing height	59	63.6	67.1	72.8
4	Sitting height	30.9	33.2	35.7	38.0
5	Seated width	12.3	12.2	17.1	15.9

a military application. For example, MIL-STD-1472 [9] states, "Design shall insure operability and maintainability by at least 90 percent of the user population" and "The design range shall include at least the 5th and 95th percentiles for design-critical body dimensions."

Furthermore, the standard states that the use of anthropometric data should take into consideration factors such as the nature and frequency of tasks to be performed, the difficulties associated with intended tasks, the position of the body during task performance, the mobility and flexibility requirements of the task, the increments in the design-critical dimensions imposed by protective garments, the need to compensate for obstacles, and so on.

Some body-related dimensions of the U.S. adult population (18 to 79 years) are presented in Table 10.1 [9–11].

Some useful pointers for engineering designers concerning the application of body force and strength are as follows [12]:

- With the use of the whole arm and shoulder, the maximum exertable force is increased.
- A person's arm strength reaches its peak around age 25.
- The maximum handgrip strength of a 25-year-old male is about 125 pounds.
- The maximum push force for side-to-side motion is about 90 pounds.
- Pull force is greater from a sitting than from a standing position.
- The degree of force that can be exerted is determined by factors such as body parts involved, direction of force applied, body position, and the object involved.

10.4 AUDITORY AND VISUAL WARNINGS IN MAINTENANCE WORK

In maintenance work various auditory and visual warning devices are used for the safety of maintenance personnel. A clear understanding of such devices is essential. Examples of warning devices used in maintenance work are sirens, bells, and buzzers.

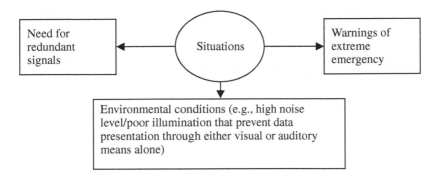

FIGURE 10.1 Situations that require the simultaneous use of both visual and auditory signals.

In maintainability design, with respect to the use of auditory warning devices, attention should be given to factors such as easy detectability, suitability to get the attention of repair personnel, use of warbling or undulating tones and sound at least 20 dB above threshold level, distinctiveness, noncontinuous and high-pitched tones above 2,000 Hz, and no requirement for interpretation when maintenance people are performing repetitive tasks [2].

Additional design recommendations for auditory warning devices to address corresponding conditions (in parentheses) are to select a frequency that makes the signal audible through other noise (presence of background noise), use low frequencies (sound is expected to pass through partitions and bend around obstacles), modulate the signal to generate intermittent beeps (signal must command maintenance person's attention), use manual shut-off mechanism (warning signal must be acknowledged), and use high intensities and avoid high frequencies (repair personnel are performing their tasks far from the signal source).

Some of the conditions for using auditory presentation are that the message is simple, the message receiving location is too brightly lit, the maintenance person is moving around continuously, the maintenance person is overburdened with visual stimuli, the message is short, and the message requires immediate action. Similarly, some of the conditions for using visual presentation are that the message is complex, the maintenance person is overburdened with auditory stimuli, the message is long, the message receiving location is too noisy, the message does not require immediate action, and the maintenance person's job allows him or her to remain in one place.

Three situations that require the simultaneous use of both visual and auditory signals are shown in Figure 10.1 [3].

10.5 HUMAN FACTORS-RELATED FORMULAS

Human factors researchers have developed many mathematical formulas for estimating human factors–related information. This section presents some of the formulas considered useful for maintainability work.

10.5.1 THE DECIBEL

The level of sound intensity is measured in term of decibels, the basic unit named after Alexander Graham Bell (1847–1922), the inventor of telephone. The sound-pressure level (SPL), in decibels, is defined by [13–14].

$$SPL = 10 \, \log_{10} \left(\frac{P^2}{P_0^2} \right) \qquad (10.1)$$

where P^2 is the sound pressure squared of the sound to be measured and P_0^2 is the standard reference sound pressure squared, representing zero decibels. Under normal conditions, P_0 is the faintest 1,000-Hz tone that an average young adult can hear.

10.5.2 CHARACTER HEIGHT ESTIMATION FORMULAS

10.5.2.1 Formula 1

Usually, for a comfortable arm reach for performing control and adjustment-oriented tasks, the instrument panels are installed at a viewing distance of 28 inches. Thus, letter, marking, and number sizes are based on this viewing distance. However, sometimes the need may arise to vary this distance; under such circumstances, the following equation can be used to estimate the required character height [13,15]:

$$H_C = H_S \, D_S \, / \, 28 \qquad (10.2)$$

where H_C is the character height estimate at the specified viewing distance (D_S) expressed in inches, and H_S is the standard or recommended character height at a viewing distance of 28 inches.

Example 10.1
A meter has to be read at a distance of 56 inches. The recommended numeral height at a viewing distance of 28 inches at low luminance is 0.31 inches. Calculate the numeral height for the viewing distance of 56 inches.

Substituting the specified data values into Equation 10.2 yields

$$H_C = (0.31)(56) / 28$$
$$= 0.62 \; inches$$

Thus, the estimate for the numeral height for the specified viewing distance is 0.62 inches.

10.5.2.2 Formula 2

This formula was developed by Peters and Adams in 1959 to determine character height by taking into consideration factors such as illumination, importance of reading accuracy, viewing distance, and viewing conditions [16]. Thus, the character height is expressed by

$$CH = \theta VD + \alpha_i + \alpha_{VC\,i} \qquad (10.3)$$

where CH is the character height in inches, VD is the viewing distance expressed in inches, α_i is the correction factor associated with importance. Its specified value for important items such as emergency labels is 0.075 and for other items is $\alpha_i = 0$. α_{VCi} is the correction factor for viewing conditions and illumination. Its recommended values for various corresponding viewing conditions and illuminations (in parentheses) are 0.06 (favorable reading conditions, above 1 foot-candle), 0.16 (unfavorable reading conditions, above 1 foot-candle), 0.16 (favorable reading conditions, below 1 foot-candle), and 0.26 (unfavorable reading conditions, below 1 foot-candle). θ is a constant whose specified value is 0.0022.

Example 10.2
Assume that the viewing distance of an instrument panel is estimated to be 42 inches. Calculate the height of the characters that should be used on the panel for $\alpha_i = 0.075$ and $\alpha_{VCi} = 0.06$.
Using the above specified values in Equation 10.3, we get

$$CH = (0.0022)(42) + 0.075 + 0.06$$
$$= 0.2274$$

Thus, the height of the characters should be 0.2274 inches.

10.5.3 Lifting Load Estimation

This formula is concerned with estimating the maximum lifting load for a person. This information could be quite useful with respect to structuring various maintenance tasks. The maximum lifting load is expressed by [17]

$$MLL = k\,(IBMS) \qquad (10.4)$$

where MLL is the maximum lifting load for a person, $IMBS$ is the isometric back muscle strength of the person, and k is a constant whose values are 1.1 and 0.95 for males and females, respectively.

10.5.4 GLARE CONSTANT ESTIMATION

For various maintainability-related tasks, glare can be a serious problem. The value
of the glare constant can be calculated by using the following equation [18]:

$$\alpha_g = \frac{\left(L_S\right)^{1.6}\left(S_a\right)^{0.8}}{\left(AVG_d\right)^2\left(L_{gb}\right)} \tag{10.5}$$

where α_g is the glare constant value, L_S is the source luminance, S_a is the solid angle
subtended at the eye by the source, AVG_d is the angle between the viewing direction
and the glare source direction, and L_{gb} is the general background luminance.

10.6 PROBLEMS

1. Write an essay on human factors in maintainability.
2. List at least fifteen typical human behaviors.
3. What are the human sensory capabilities? Discuss at least three such
   capabilities in detail.
4. List at least five useful pointers for engineering designers concerning the
   application of body force and strength.
5. List factors to which attention should be given in maintainability design
   in regard to the use of auditory warning devices.
6. List important conditions for using auditory presentation.
7. List important conditions for using visual presentation.
8. Define decibel (dB).
9. The estimated viewing distance of an instrument panel is 50 inches.
   Calculate the height of the characters that should be used on the panel if
   the values of the importance correction factor and the illumination and
   viewing conditions correction factor are 0.075 and 0.16, respectively.
10. Write down the formula for estimating the maximum lifting load for a
    person.

REFERENCES

1. Chapanis, A., *Man-Machine Engineering*, Wadsworth, Belmont, CA, 1965.
2. Dhillon, B.S., *Engineering Maintainability*, Gulf Publishing, Houston, TX, 1999.
3. *Engineering Design Handbook: Maintainability Guide for Design*, AMCP 706-134,
   Department of Defense, Washington, DC, 1972.
4. Woodson, W.E., *Human Factors Design Handbook*, McGraw-Hill, New York, 1981.
5. Nertney, R.J., Bullock, M.G., *Human Factors in Design*, Report No. ERDA-76-45-2,
   The Energy Research and Development Administration, U.S. Department of Energy,
   Washington, DC, 1976.
6. *Engineering Design Handbook: Maintainability Engineering Theory and Practice*,
   AMCP, 706-133, Department of Defense, Washington, DC, 1976.

7. Lederman, S., Heightening tactile impressions of surface texture, in *Active Touch*, Gordon, G., Ed., Pergamon Press, Elmsford, NY, 1978, pp. 20–32.
8. Woodson, W., Human engineering suggestions for designers of electronic equipment, in *NEL Reliability Design Handbook*, U.S. Naval Electronics Laboratory, San Diego, CA, 1955, pp. 12.1–12.5.
9. *Human Engineering Design for Military Systems, Equipment, and Facilities*, MIL-STD-1472, Department of Defense, Washington, DC, 1972.
10. Woodson, W.E., *Human Factors Design Handbook*, McGraw-Hill, New York, 1981.
11. Dhillon, B.S., *Advanced Design Concepts for Engineers*, Technomic, Lancaster, PA, 1998.
12. Henney, K., Ed., *Reliability Factors for Ground Electronic Equipment*, The Rome Air Development Center, Griffis Air Force Base, Rome, NY, 1955.
13. McCormick, E.J. and Sanders, M.S., *Human Factors in Engineering and Design*, McGraw-Hill, New York, 1982.
14. Adams, J.A., *Human Factors Engineering*, MacMillan, New York, 1989.
15. Dale Huchingson, R., *New Horizons for Human Factors in Design*, McGraw-Hill, New York, 1981.
16. Peters, G.A. and Adams, B.B., Three criteria for readable panel markings, *Product Engineering*, 30(2), 55–57, 1959.
17. Poulsen, E., Jorgensen, C., Back muscle strength, lifting and stooped working postures, *Applied Ergonomics,* Vol. 2, 1971, pp. 133–137.
18. Oborne, D.J., *Ergonomics at Work*, John Wiley & Sons, New York, 1982.

11 Introduction to Engineering Maintenance

11.1 NEED FOR MAINTENANCE

Each year billons of dollars are spent on engineering equipment maintenance world-wide, and today's maintenance practices are market driven, in particular for the manufacturing and process industry, service suppliers, and so on [1]. Because of this, there is a definite need for effective asset management and maintenance practices that can positively influence success factors such as price, profitability, quality, reliable delivery, safety, and speed of innovation.

In the future engineering equipment will be even more computerized and complex. Further computerization of equipment will increase the importance of software maintenance significantly, approaching, if not equaling hardware maintenance. In addition, factors such as increased computerization and complexity will result in greater emphasis on maintenance activities with respect to areas such as cost effectiveness, quality, safety, and human factors [2]. In the future creative thinking and new strategies will definitely be required to realize all potential benefits and turn them into profitability.

11.2 FACTS AND FIGURES RELATED TO ENGINEERING MAINTENANCE

Some of the facts and figures concerning engineering maintenance are as follows:

- U.S. industry spends over $300 billion annually on plant maintenance and operations [3].
- It is estimated that the cost of maintaining a military jet aircraft is approximately $1.6 million per year, and about 11% of the operating cost for an aircraft accounts for maintenance activities [4].
- Over the years the size of a plant maintenance group in a manufacturing organization has varied from 5 to 10% of the total operating force [5] — 1 to 17 persons in 1969 and 1 to 12 persons in 1981 [5].
- For fiscal year 1997, the request of the U.S. Department of Defense for their operation and maintenance budget was $79 billion [6].

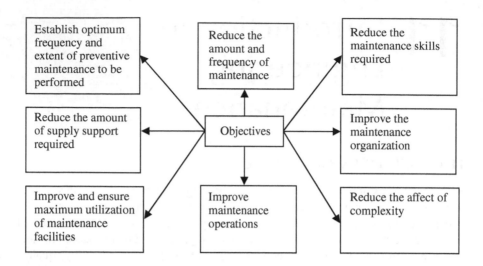

FIGURE 11.1 Important objectives of maintenance engineering.

- In 1970 a British Ministry of Technology Working Party document reported that the United Kingdom annual maintenance cost was around 3000 million pounds [7,8].
- The U.S. Department of Defense spends approximately $12 billion annually on depot maintenance of weapon systems and equipment [9].

11.3 MAINTENANCE ENGINEERING OBJECTIVES

There are many objectives of maintenance engineering. Some of the important ones are shown in Figure 11.1 [10].

11.4 MAINTENANCE-RELATED DATA INFORMATION SOURCES

There are many sources for obtaining maintenance-related data. Table 11.1 presents five such sources.

11.5 MAINTENANCE MEASURES

Many indexes to measure maintenance activity performance have been developed. Usually, the values of these indexes are calculated periodically to monitor their trends or compare them with established standard values. This section presents some of these indexes [5,11–13].

TABLE 11.1
Sources for Obtaining Maintenance-Related Data

No.	Source Name and Address
1	National Technical Information Service (NTIS), 5285 Port Royal Road, Springfield, VA
2	GIDEP Data, Government Industry Data Exchange Program (GIDEP) Operations Center, Fleet Missile Systems, Analysis, and Evaluation Group, Department of Navy, Corona, CA
3	Reliability Analysis Center, Rome Air Development Center, Griffis Air Force Base, Rome, NY
4	Defense Technical Information Service, DTIC-FDAC, 8725 John J. Kingman Road, Suite 0944, Fort Belvoir, VA
5	Data on Trucks and Vans, Commanding General, Attn: DRSTA-QRA, U.S. Army Automotive-Tank Command, Warren, MI

11.5.1 INDEX 1

This index relates the maintenance cost to the total investment in plant and equipment and is defined by

$$I_{mi} = \frac{C_m}{C_i} \tag{11.1}$$

where I_{mi} is the index parameter, C_m is the total maintenance cost, and C_i is the total amount of investment in plant and equipment.

The approximate average values for this index in the chemical and steel industries are 3.8 and 8.6%, respectively.

11.5.2 INDEX 2

This index measures the maintenance budget plan accuracy and is defined as follows:

$$I_{ab} = \frac{C_{am}}{C_{bm}} \tag{11.2}$$

where I_{ab} is the index parameter, C_{bm} is the budgeted maintenance cost, and C_{am} is the actual maintenance cost.

Large variances in the values of this index indicate the need for immediate attention.

11.5.3 INDEX 3

This index relates the maintenance cost to the total sales revenue and is defined as follows:

$$I_{ms} = \frac{C_m}{C_{Sr}} \tag{11.3}$$

where I_{ms} is the index parameter and C_{Sr} is the total sales revenue.

According to some documents, the average expenditure of the maintenance activity for all industry is around 5% of sales revenue. However, there is a wide variation among industries. For example, the expenditure for the maintenance activity for chemical and steel industries is around 6.8% and 12.8% of sales revenue, respectively.

11.5.4 INDEX 4

This index relates maintenance cost to total man-hours worked and is defined as follows:

$$I_{mm} = \frac{C_m}{C_{mh}}$$
(11.4)

where I_{mm} is the index parameter and C_{mh} is the total number of man-hours worked.

11.5.5 INDEX 5

This index relates the total maintenance cost to the total output (i.e., in units such as tons and megawatts) by the organization in question and is defined by

$$I_{mo} = \frac{C_m}{T_o}$$
(11.5)

where I_{mo} is the index parameter and T_o is the total output by the organization in question, expressed in units such as tons, megawatts, and gallons.

11.5.6 INDEX 6

This index relates the total maintenance cost to the total manufacturing cost and is defined as follows:

$$I_{mtm} = \frac{C_m}{C_{tm}}$$
(11.6)

where I_{mtm} is the index parameter and C_{tm} is the total manufacturing cost.

11.5.7 INDEX 7

This index is quite useful for measuring inspection effectiveness and is defined as follows:

$$I_{jc} = \frac{TNJRI}{TNIC}$$
(11.7)

where I_{jc} is the index parameter, $TNIC$ is the total number of inspections completed, and $TNJRI$ is the total number of jobs resulting from inspections.

11.5.8 INDEX 8

This is an important index used to measure maintenance effectiveness with respect to man-hours associated with emergency and unscheduled jobs and total maintenance man-hours worked. The index is defined by

$$I_{mh} = \frac{TMH_{eu}}{TMH_w} \qquad (11.8)$$

where I_{mh} is the index parameter, TMH_w is the total number of maintenance man-hours worked, and TMH_{eu} is the total number of man-hours associated with emergency and unscheduled jobs.

11.5.9 INDEX 9

This index is often used in material control areas and is defined as follows:

$$I_{mc} = \frac{THJ_{pam}}{TJ_p} \qquad (11.9)$$

where I_{mc} is the index parameter, TJ_{pam} is the total number of planned jobs awaiting material, and TJ_p is the total number of planned jobs.

11.6 SAFETY IN MAINTENANCE

Safety in maintenance is becoming an important issue, as accidents occurring during maintenance work or concerning maintenance are increasing significantly. For example, in 1994 around 13.61% of all accidents in the U.S. mining industry occurred during maintenance work and they have been increasing at a significant rate annually since 1990 [14,15].

Some of the main reasons for safety problems in maintenance are poor safety standards and tools, poor equipment design, poor training of maintenance personnel, insufficient time to perform required maintenance tasks, poorly written maintenance instructions and procedures, poor management, poor work environments, and inadequate work tools [15].

One of the important ways to improve maintenance safety is to reduce the requirement for maintenance as much as possible in products and systems during their design phase. When the need for maintenance cannot be avoided, designers should follow guidelines such as those listed below for improving safety in maintenance [16]:

- Eradicate the need for performing maintenance and adjustments close to hazardous operating parts or equipment.
- Keep design as simple as possible, because complexity usually adds to maintenance problems.
- Aim to eliminate the requirement for special tools or equipment.
- Provide appropriate guards against moving articles or parts and interlocks for blocking accesses to hazardous locations.

- Incorporate appropriate fail-safe designs for preventing injury and damage if a failure occurs.
- Design for easy accessibility so that items requiring maintenance can easily be checked, removed, replaced, or serviced.
- Develop designs or procedures that minimize the occurrence of maintenance errors.
- Develop the design in a manner that reduces the probability of maintenance personnel being injured by escaping high-pressure gas, electric shock, contact with a hot surface, and so on.
- Incorporate appropriate devices or other measures for early prediction and detection of all potential failures so that the required maintenance can be performed prior to failure with somewhat reduced risk of hazard.

11.7 QUALITY IN MAINTENANCE

Maintenance quality provides some degree of confidence that repaired or maintained items will function safely and reliably [17,18]. Quality in maintenance is very important because poor quality maintenance can lead to severe consequences. For example, the following three tragedies are believed to be, directly or indirectly, the result of poor quality maintenance [15]:

- In 1986 the space shuttle *Challenger* exploded and all seven crew members lost their lives [17–18]. A subsequent investigation identified the cause of the disaster as the failure of the pressure seal in the aft field joint of the right solid rocket motor. Furthermore, the investigation concluded that a high-quality maintenance program would have successfully tracked and discovered the cause of the disaster.
- In 1990 10 people died as the result of a serious steam leak in the fire room on the *U.S.S. Iwo Jima* (LPH2), a U.S. Navy ship [19]. Failure of the service turbine generator root-valve bonnet fastener was identified as the main cause of this tragedy. Further investigation revealed that ship's personnel furnished the replacement fasteners without properly verifying if the requirements of the technical manual and drawings were fully satisfied.
- In 1963 the *U.S.S. Thresher*, a U.S. Navy nuclear submarine, was lost at sea because of flooding in its engine room [17,19]. An investigation identified a piping failure in one of the salt water systems as the most likely cause for the disaster. Consequently, many changes were recommended in the submarine design and maintenance processes.

Past experiences indicate that postmaintenance testing (PMT) is quite useful for increasing the quality of maintenance. Its three main objectives are (a) to ensure that no new deficiencies have been introduced, (b) to ensure that the original deficiency has been eradicated properly, and (c) to ensure that the item in question is ready to carry out its stated mission [20]. In order to increase the quality of maintenance, PMT should not only be performed after all corrective maintenance activities but also after some preventive maintenance activities, as considered appropriate.

11.8 PROBLEMS

1. Discuss the need for maintenance.
2. Discuss at least five facts and figures concerning engineering maintenance.
3. What are the important objectives of maintenance engineering?
4. List at least three maintenance-related data information sources.
5. In your opinion, what is the most important maintenance index or measure?
6. Discuss two general maintenance indexes.
7. Discuss the importance of safety in engineering maintenance.
8. Discuss the need for quality in maintenance activities.
9. List at least nine useful guidelines for equipment designers to improve safety in the maintenance activity.
10. What are the important causes of safety problems in engineering maintenance.

REFERENCES

1. Zweekhorst, A., Evolution of maintenance, *Maintenance Technology*, October, 9–14, 1996.
2. Tesdahl, S.A. and Tomlingson, P.D., Equipment management breakthrough maintenance strategy for the 21st century, *Proceedings of the First International Conference on Information Technologies in the Minerals Industry*, December 1997, pp. 39–58.
3. Latino, C.J., *Hidden Treasure: Eliminating Chronic Failures Can Cut Maintenance Costs up to 60%*, Reliability Center, Hopewell, VA, 1999.
4. Kumar, V.D., New trend in aircraft reliability and maintenance measures, *Journal of Quality in Maintenance Engineering*, 5(4), 287–299, 1999.
5. Niebel, B.W., *Engineering Maintenance Management*, Marcel Dekker, New York, 1994.
6. *1997 DOD Budget: Potential Reductions to Operation and Maintenance Program*, U.S. General Accounting Office, Washington, DC, 1996.
7. Kelly, A., *Management of Industrial Maintenance*, Newes-Butterworths, London, 1978.
8. *Report by the Working Party on Maintenance Engineering*, Department of Industry, London, 1970.
9. *Report on Infrastructure and Logistics*, Department of Defense, Washington, DC, 1995.
10. *Engineering Design Handbook: Maintenance Engineering Techniques*, AMCP 706-132, Department of the Army, Washington, DC, 1975.
11. Westerkemp, T.A., *Maintenance Manager's Standard Manual*, Prentice Hall, Paramus, NJ, 1997.
12. Hartmann, E., Knapp, D.J., Johnstone, J.J., and Ward, K.G., *How to Manage Maintenance*, American Management Association, New York, 1994.
13. Stoneham, D., *The Maintenance Management and Technology Handbook*, Elsevier Science, Oxford, U.K., 1998.
14. *Accidents Facts*, National Safety Council, Chicago, IL, 1999.

15. Dhillon, B.S., *Engineering Maintenance: A Modern Approach*, CRC Press, Boca Raton, FL, 2002.
16. Hammer, W., *Product Safety Management and Engineering*, Prentice Hall, Englewood Cliffs, NJ, 1980.
17. *Joint Fleet Maintenance Manual*, Vol. 5, *Quality Maintenance, Submarine Maintenance Engineering*, U.S. Navy, Portsmouth, NH, 1989.
18. *Report: The Presidential Commission on the Space Shuttle Challenger*, Vol. 1, Washington, DC, 1986.
19. Elsayed, E.A., *Reliability Engineering*, Addison Wesley Longman, Reading, MA, 1996.
20. *Guidelines to Good Practices for Post-Maintenance Testing at DOE Nuclear Facilities*, DOE-STD-1065-94, Department of Energy, Washington, DC, 1994.

12 Corrective and Preventive Maintenance

12.1 INTRODUCTION

Corrective maintenance is the remedial action performed because of failure or deficiencies found during preventive maintenance or otherwise, to repair an item to its operating state [1–4]. Normally, corrective maintenance is an unplanned maintenance action that requires urgent attention that must be added, integrated with, or substituted for previously scheduled work. Corrective maintenance or repair is an important element of overall maintenance activity.

Preventive maintenance is an important element of a maintenance activity and within a maintenance department it normally accounts for a significant proportion of the overall maintenance activity. Preventive maintenance is the care and servicing by maintenance personnel to keep facilities in a satisfactory operational state by providing for systematic inspection, detection, and correction of incipient failures either before their development into major failures or before their occurrence [2,4]. There are many objectives of performing preventive maintenance including improving capital equipment's productive life, reducing production losses caused by equipment failure, minimizing critical equipment breakdowns, and improving the health and safety of maintenance personnel [5].

This chapter presents various important aspects of both corrective maintenance and preventive maintenance.

12.2 TYPES OF CORRECTIVE MAINTENANCE

Corrective maintenance may be grouped under the following five categories [2,4,6]:

Fail repair: This is concerned with restoring the failed item or equipment to its operational state.

Overhaul: This is concerned with repairing or restoring an item or equipment to its complete serviceable state meeting requirements outlined in maintenance serviceability standards, using the "inspect and repair only as appropriate" method.

Salvage: This is concerned with the disposal of nonrepairable materials and utilization of salvaged materials from items that cannot be repaired in the overhaul, repair, or rebuild programs.

Servicing: This type of corrective maintenance may be required because of a corrective maintenance action; for example, engine repair can result in requirement for crankcase refill, welding on, and so on.

Rebuild: This is concerned with restoring an item or equipment to a standard as close as possible to its original state with respect to appearance, performance, and life expectancy. This is accomplished through actions such as complete disassembly, examination of all parts, replacement or repair of unserviceable or worn components according to original specifications and manufacturing tolerances, and reassembly and testing to original production requirements.

12.3 CORRECTIVE MAINTENANCE STEPS, DOWNTIME COMPONENTS, AND TIME-REDUCTION STRATEGIES AT SYSTEM LEVEL

Different authors and researchers have proposed different steps for carrying out corrective maintenance [1,3]. For our purpose, we assume that corrective maintenance can be performed in the following five steps [4]:

Failure recognition: Recognizing the existence of a failure
Failure localization: Localizing the failure within the system to a specific piece of equipment item
Diagnosis within the equipment or item: Diagnosis within an item or equipment to identify specific failed part or component.
Failed part replacement or repair: Replacing or repairing failed parts or components.
Return system to service: Checking out and returning the system back to service.

Corrective maintenance downtime is made up of three major components as shown in Figure 12.1 [4,7].

Active repair time is made up of six subcomponents: checkout time, preparation time, fault correction time, fault location time, adjustment and calibration time, and spare item obtainment time [4,7].

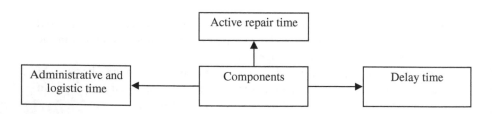

FIGURE 12.1 Major corrective maintenance downtime components.

In order to improve corrective maintenance effectiveness, it is important to reduce corrective maintenance time. Some of the useful strategies for reducing system-level corrective maintenance time are [2,8]:

Improve accessibility: Past experiences indicate that often a significant amount of time is spent accessing failed parts. Careful attention to accessibility during design can help to lower the accessibility time of parts and, consequently, the corrective maintenance time.

Improve interchangeability: Effective functional and physical interchangeability is an important factor in removing and replacing parts or components, thus lowering corrective maintenance time.

Improve fault recognition, location, and isolation: Past experiences indicate that within a corrective maintenance activity, fault recognition, location, and isolation consume the most time. Factors that help to reduce corrective maintenance time are good maintenance procedures, well-trained maintenance personnel, well-designed fault indicators, and unambiguous fault isolation capability.

Consider human factors: During design, paying careful attention to human factors such as selection and placement of indicators and dials; size, shape, and weight of components; readability of instructions; information processing aids; and size and placement of access and gates can help lower corrective maintenance time significantly.

Employ redundancy: This is concerned with designing in redundant parts or components that can be switched in during the repair of faulty parts so that the equipment or system continues to operate. In this case, although the overall maintenance workload may not be reduced, the downtime of the equipment could be impacted significantly.

12.4 CORRECTIVE MAINTENANCE MEASURES

There are many corrective maintenance-related measures. Two of those measures are presented below [4,8,9].

12.4.1 MEAN CORRECTIVE MAINTENANCE TIME

This is an important measure of corrective maintenance and is defined by

$$CMMT = \frac{\sum \lambda_i \, CMT_i}{\sum \lambda_i} \tag{12.1}$$

where $CMMT$ is the mean corrective maintenance time, λ_i is the failure rate of the ith equipment element, and CMT_i is the corrective maintenance time of the ith equipment element.

Usually corrective maintenance times are described by normal, lognormal, and exponential probability distributions. Examples of the types of equipment that follow these distributions are:

Normal distribution: Corrective maintenance times of mechanical or electromechanical equipment with a remove and replacement maintenance concept often follow this distribution.

Lognormal distribution: Corrective maintenance times of electronic equipment that does not possess built-in test capability usually follow this distribution.

Exponential distribution: Corrective maintenance times of electronic equipment with a good built-in test capability and rapid remove and replace maintenance concept often follow this distribution.

12.4.2 MEDIAN ACTIVE CORRECTIVE MAINTENANCE TIME

This is another important measure of corrective maintenance. It usually provides the best average location of the sample data and is the 50th percentile of all values of corrective maintenance time. Median active corrective maintenance time is a measure of the time within which 50% of all corrective maintenance activities can be performed. The computation of this measure is subject to the probability distribution describing corrective maintenance times. Thus, the median of corrective maintenance times following a lognormal distribution is expressed by [2,8]

$$MACMT = anti \log \left[\frac{\sum \lambda_i \log CMT_i}{\sum \lambda_i} \right] \tag{12.2}$$

where *MACMT* is the median active corrective maintenance time.

12.5 MATHEMATICAL MODELS FOR PERFORMING CORRECTIVE MAINTENANCE

Many mathematical models are available in the published literature that can be used in performing corrective maintenance. This section presents two such models. These models take into consideration item failure and corrective maintenance rates and can be used to predict item, equipment, and system availability, reliability, probability of being in a failed state (i.e., undergoing repair or corrective maintenance), mean time to failure, and so on.

12.5.1 MATHEMATICAL MODEL 1

This model represents a system that can be in either operating or failed state. The failed system is repaired back to its operating state. Most industrial systems, equipment, and items follow this pattern. The system-state space diagram is shown

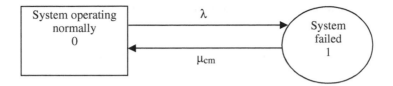

FIGURE 12.2 System–state space diagram.

in Figure 12.2 [9]. The numerals in the box and circle in Figure 12.2 denote system states. The following assumptions are associated with the model:

- All system failures are statistically independent.
- System failure and repair (i.e., corrective maintenance) rates are constant.
- The repaired system is as good as new.

The symbols used to develop equations for the model are defined below:

- λ is the system failure rate.
- μ_{cm} is the system corrective maintenance or repair rate.
- j is the jth system state; $j = 0$ (system operating normally), $j = 1$ (system failed).
- $P_j(t)$ is the probability that the system is in state j at time t for $j = 0$ and $j = 1$.

With the aid of the Markov method presented in Chapter 4 and Figure 12.2, we write down the following two equations [2,10]:

$$\frac{dP_0(t)}{dt} + \lambda P_0(t) = \mu_{cm} P_1(t) \tag{12.3}$$

$$\frac{dP_1(t)}{dt} + \mu_{cm} P_1(t) = \lambda P_0(t) \tag{12.4}$$

At time $t = 0$, $P_0(0) = 1$ and $P_1(0) = 0$.

Solving Equation 12.3 and Equation 12.4, we get

$$P_0(t) = \frac{\mu_{cm}}{\lambda + \mu_{cm}} + \frac{\lambda}{\lambda + \mu_{cm}} e^{-(\lambda + \mu_{cm})t} \tag{12.5}$$

and

$$P_1(t) = \frac{\lambda}{\lambda + \mu_{cm}} - \frac{\lambda}{\lambda + \mu_{cm}} e^{-(\lambda + \mu_{cm})t} \tag{12.6}$$

The system availability and unavailability are given by

$$AV(t) = P_0(t) = \frac{\mu_{cm}}{\lambda + \mu_{cm}} + \frac{\lambda}{\lambda + \mu_{cm}} e^{-(\lambda + \mu_{cm})t} \tag{12.7}$$

and

$$UA(t) = P_1(t) = \frac{\lambda}{\lambda + \mu_{cm}} - \frac{\lambda}{\lambda + \mu_{cm}} e^{-(\lambda + \mu_{cm})t} \tag{12.8}$$

where $AV(t)$ is the system availability at time t and $UA(t)$ is the system unavailability at time t.

As t becomes very large, Equation 12.7 and Equation 12.8 reduce to

$$AV = \frac{\mu_{cm}}{\lambda + \mu_{cm}} \tag{12.9}$$

and

$$UA = \frac{\lambda}{\lambda + \mu_{cm}} \tag{12.10}$$

where AV is the system steady-state availability, and UA is the system steady-state unavailability.

Since $\mu_{cm} = 1/CMMT$ and $\lambda = 1/MTTF$, Equation 12.9 and Equation 12.10 become

$$AV = \frac{MTTF}{CMMT + MTTF} \tag{12.11}$$

and

$$UA = \frac{CMMT}{CMMT + MTTF} \tag{12.12}$$

where $MTTF$ is the system mean time to failure.

Example 12.1

A system's mean time to failure is 2,000 hours and its mean corrective maintenance time, or mean time to repair, is 25 hours. Calculate the system steady-state unavailability if the system failure and corrective maintenance time follow exponential distribution.

Using the specified data values in Equation 12.12 yields

$$UA = \frac{25}{25 + 2,000} = 0.0123$$

This means the system steady-state unavailability is 0.0123, or there is 1.23% chance that the system will be unavailable for service.

12.5.2 Mathematical Model 2

This model represents a parallel system made up of two identical units. For system success, at least one unit must operate normally. The system fails when both the units fail. Repair or corrective maintenance begins as soon as a unit fails to return to its operating state. The system-state space diagram is shown in Figure 12.3. The numerals in boxes and circle denote system states. The model is subject to the following assumptions:

- Unit failure and repair or corrective maintenance rates are constant.
- The system contains two independent and identical units.
- No repair or corrective maintenance is performed when both the units fail or the system fails.
- The repaired unit is as good as new.

The symbols used to develop equations for two models are defined below:

- λ is the unit failure rate.
- μ is the unit repair or corrective maintenance rate.
- j is the jth system state; $j = 0$ (both units are working normally), $j = 1$ (one unit failed, the other operating normally), $j = 2$ (both units failed).
- $P_j(t)$ is the probability that the system is in state j at time t, for $j = 0, 1, 2$.

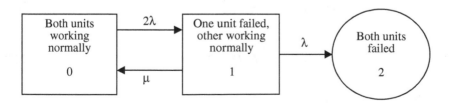

FIGURE 12.3 The two-unit parallel system-state–space diagram.

Using the Markov method and Figure 12.3, we get the following equations [2,10,11]:

$$\frac{dP_0(t)}{dt} + 2\lambda P_0(t) = \mu \, P_1(t) \tag{12.13}$$

$$\frac{dP_1(t)}{dt} + (\mu + \lambda) \, P_1(t) = 2\lambda P_0(t) \tag{12.14}$$

$$\frac{dP_2(t)}{dt} = \lambda P_1(t) \tag{12.15}$$

At time $t = 0$, $P_0(0) = 1$ and $P_1(0) = P_2(0) = 0$.

Solving Equation 12.13 to Equation 12.15, we get

$$P_0(t) = \left[\frac{\lambda + \mu + D_1}{D_1 - D_2}\right] e^{D_1 t} - \left[\frac{\lambda + \mu + D_2}{D_1 - D_2}\right] e^{D_2 t} \tag{12.16}$$

$$P_1(t) = \left[\frac{2\lambda}{D_1 - D_2}\right] e^{D_1 t} - \left[\frac{2\lambda}{D_1 - D_2}\right] e^{D_2 t} \tag{12.17}$$

and

$$P_2(t) = 1 + \left[\frac{D_2}{D_1 - D_2}\right] e^{D_1 t} - \left[\frac{D_1}{D_1 - D_2}\right] e^{D_2 t} \tag{12.18}$$

where

$$D_1, D_2 = \frac{-(3\lambda + \mu) \pm \left[(3\lambda + \mu)^2 - 8\lambda^2\right]^{1/2}}{2} \tag{12.19}$$

$$D_1 D_2 = 2\lambda^2 \tag{12.20}$$

and

$$D_1 + D_2 = -(3\lambda + \mu) \tag{12.21}$$

The parallel system reliability with repair is given by

$$R_{ps}(t) = P_0(t) + P_1(t)$$ (12.22)

where $R_{ps}(t)$ is the parallel system reliability with repair at time t.

The system mean time to failure with repair is given by

$$MTTF_{ps} = \int_0^\infty R_{ps}(t)\,dt$$

$$= \frac{3\lambda + \mu}{2\lambda^2}$$ (12.23)

where $MTTF_{ps}$ is the parallel system mean time to failure with repair.

Since $\lambda = 1/MTTF$ and $\mu = 1/MTTR$, Equation 12.23 becomes

$$MTTF_{ps} = \frac{MTTF}{2MTTR}\left[3\,MTTR + MTTF\right]$$ (12.24)

where MTTF is the unit mean time to failure and $MTTR$ is the unit mean time to repair, or the mean corrective maintenance time.

Example 12.2

An engineering system is composed of two independent and identical units, and at least one of the units must operate normally for system success. Both the units form a parallel configuration. A failed unit is repaired, but the failed system is never repaired. The unit times to failure and repair (i.e., corrective maintenance) are exponentially distributed.

The unit mean time to failure and mean time to repair are 200 hours and 10 hours, respectively. Calculate the system mean time to failure with and without the performance of corrective maintenance and comment on the end results.

Using the data values in Equation 12.24 yields

$$MTTF_{ps} = \frac{200}{2(10)}\left[3(10) + 200\right]$$

$$= 2,300\ hours$$

Setting $\mu = 0$ and inserting the specified data value into Equation 12.23 yields

$$MTTF_{ps} = \frac{3}{2\lambda} = \frac{3\,MTTF}{2} = \frac{3(200)}{2}$$

$$= 300\ hours$$

Thus, the system mean time to failure with and without the performance of corrective maintenance are 2,300 hours and 300 hours, respectively. This means the performance of corrective maintenance or repair on a unit has helped increase system mean time to failure from 300 hours to 2,300 hours.

12.6 PREVENTIVE MAINTENANCE COMPONENTS AND PRINCIPLE FOR CHOOSING ITEMS FOR PREVENTIVE MAINTENANCE

There are seven elements of preventive maintenance [2,4]:

- **Inspection:** Periodically inspecting items to determine their serviceability by comparing their physical, mechanical, electrical, and other characteristics to established standards
- **Calibration:** Detecting and adjusting any discrepancy in the accuracy of the material or parameter being compared to the established standard value
- **Testing:** Periodically testing to determine serviceability and detect mechanical or electrical degradation
- **Adjustment:** Periodically making adjustments to specified variable elements to achieve optimum performance
- **Servicing:** Periodically lubricating, charging, cleaning, and so on, materials or items to prevent the occurrence of incipient failures
- **Installation:** Periodically replacing limited-life items or items experiencing time cycle or wear degradation to maintain the specified tolerance level
- **Alignment:** Making changes to an item's specified variable elements to achieve optimum performance

The following formula principle can be quite useful in deciding whether to implement a preventive maintenance program for an item or system [12,13]:

$$(n)(C_a)(\theta) > C_{pm} \tag{12.25}$$

where n is the total number of breakdowns, θ is 70% of the total cost of breakdowns, C_a is the average cost per breakdown, and C_{pm} is the total cost of the preventive maintenance system.

12.7 STEPS FOR DEVELOPING PREVENTIVE MAINTENANCE PROGRAM

Development of an effective preventive maintenance program requires the availability of items such as test instruments and tools, accurate historical records of equipment, skilled personnel, service manuals, manufacturer's recommendations, past data from similar equipment, and management support and user cooperation [14].

A highly effective preventive maintenance program can be developed in a short time by following the steps listed below [15]:

- **Identify and select the areas:** Identify and select of one or two important areas on which to concentrate the initial preventive maintenance effort. The main objective of this step is to obtain good results in areas that are highly visible.
- **Highlight the preventive maintenance requirements:** Define the preventive maintenance needs and then develop a schedule for two types of tasks: daily preventive maintenance inspections and periodic preventive maintenance assignments.
- **Determine assignment frequency:** Establish the frequency of assignments and review the item or equipment records and conditions. The frequency depends on factors such as vendor recommendations, the experience of personnel familiar with the equipment or item under consideration, and recommendations from engineers.
- **Prepare the preventive maintenance assignments:** Prepare the daily and periodic assignments in an effective manner and then get them approved.
- **Schedule the preventive maintenance assignments:** Schedule the defined preventive maintenance assignments on the basis of a 12-month period.
- **Expand the preventive maintenance program as appropriate:** Expand the preventive maintenance program to other areas on the basis of experience gained from the pilot preventive maintenance projects.

12.8 PREVENTIVE MAINTENANCE MEASURES

There are many preventive maintenance-related measures. This section presents two such measures taken from the published literature [2,4,8].

12.8.1 Mean Preventive Maintenance Time

This is an important measure of preventive maintenance. It is the average equipment downtime required to perform scheduled preventive maintenance. Mean preventive maintenance time is expressed by

$$PMT_m = \left[\sum_{j=1}^{k} f_j \, PMT_{mj} \right] / \sum_{j=1}^{k} f_j \qquad (12.26)$$

where PMT_m is the mean preventive maintenance time; k is the total number of data points; PMT_{mj} is the average time required to carry out j preventive maintenance tasks for $j = 1, 2, 3, \ldots, k$; and f_j is the frequency of j preventive maintenance task in tasks per operating hour after adjustment for item or equipment duty cycle.

12.8.2 MEDIAN PREVENTIVE MAINTENANCE TIME

This is another important measure of preventive maintenance. Median preventive maintenance time is the equipment downtime required to perform 50% of all scheduled preventive maintenance actions under the conditions stated for median preventive maintenance time. For lognormal distributed preventive maintenance times, the median preventive maintenance time is defined by

$$MPT_m = anti \log \left[\frac{\sum_{j=1}^{k} \lambda_j \ \log \ PMT_{mj}}{\sum_{j=1}^{k} \lambda_j} \right] \tag{12.27}$$

where MPT_m is the median preventive maintenance time and λ_j is the constant failure rate of component j of the equipment for which maintainability is to be determined, adjusted for factors such as tolerance and interaction failures, duty cycle, and catastrophic failures that will result in deterioration of equipment performance to the degree that a maintenance-related action will be taken for $j = 1, 2, 3, ..., k$.

12.9 MATHEMATICAL MODELS FOR PERFORMING PREVENTIVE MAINTENANCE

Many mathematical models have been developed to perform various types of preventive maintenance. This section presents two such models [2,16,17].

12.9.1 MODEL 1

Inspections are an important component of preventive maintenance. Usually, inspections are disruptive, but they reduce equipment downtime because they reduce failures. This model is concerned with obtaining the optimum number of inspections per facility per unit of time. Total facility downtime is expressed by

$$TFDT = \frac{\left(DT_f\right)c}{x} + x\left(DT_i\right) \tag{12.28}$$

where $TFDT$ is the total downtime per unit of time for a given facility, x is the number of inspections per facility per unit of time, DT_i is the facility downtime per inspection, DT_f is the facility downtime per failure or breakdown, and c is a constant associated with a specific facility.

By differentiating Equation 12.28 with respect to x, we obtain

$$\frac{d\,TFDT}{dx} = -\left[\frac{c\left(DT_f\right)}{x^2} \right] + DT_i \tag{12.29}$$

By equating Equation 12.29 to zero and then rearranging it, we get

$$x^* = \left[\frac{c\left(DT_f\right)}{DT_i} \right]^{1/2}$$
(12.30)

where x^* is the optimum number of inspections per facility per unit of time. Inserting Equation 12.30 into Equation 12.28 yields

$$TFDT^* = 2\left[CDT_f \, DT_i \right]^{1/2}$$
(12.31)

where $TFDT^*$ is the total optimal downtime per unit of time for a facility.

Example 12.3

The following data values are associated with an engineering facility:

- $c = 3$
- $DT_i = 0.03$ month
- $DT_f = 0.2$ month

Calculate the optimal number of inspections per month by using Equation 12.30. Substituting the given data values into Equation 12.30, we get

$$x^* = \left[\frac{(3)(0.2)}{0.03} \right]^{1/2}$$
$$= 4.47 \; inspections \; per \; month$$

Thus, the approximate number of monthly optimal inspections is 4.

12.9.2 MODEL 2

This is another useful mathematical model that represents a system that can either undergo periodic preventive maintenance or fail completely. The failed system is repaired. The system-state space diagram is shown in Figure 12.4 [18].

This model can predict items such as system availability, probability of system failure, and probability of the system being down for preventive maintenance. The model is subject to the following assumptions:

- System failure, repair, and preventive maintenance rates are constant.
- After preventive maintenance or repair the system is as good as new.

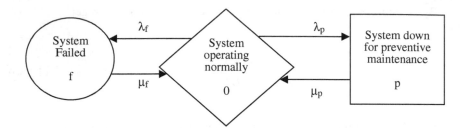

FIGURE 12.4 State–space diagram for model 2.

The symbols used to develop equations for the model are defined below:

- λ_p is the rate of the system being down for preventive maintenance.
- λ_f is the system failure rate.
- μ_p is the rate of system preventive maintenance performance.
- μ_f is the system repair or corrective maintenance rate.
- i is the ith system state; $i = 0$ (system operating normally), $i = p$ (system down for preventive maintenance), $i = f$ (system failed).
- $P_i(t)$ is the probability that the system is in state i at time t for $i = 0, p, f$.

Using the Markov method and Figure 12.4, we get the following equations [19]:

$$\frac{dP_0(t)}{dt} + \left(\lambda_f + \lambda_p\right) P_0(t) = \mu_p P_p(t) + \mu_f P_f(t) \tag{12.32}$$

$$\frac{dP_f(t)}{dt} + \mu_f P_f(t) = \lambda_f P_0(t) \tag{12.33}$$

$$\frac{dP_p(t)}{dt} + \mu_p P_p(t) = \lambda_p P_0(t) \tag{12.34}$$

At time $t = 0$, $P_0(0) = 1$, $P_f(0) = 0$, and $P_p(0) = 0$.
Solving Equation 12.32 to Equation 12.34, we get

$$P_0(t) = \frac{\mu_f \mu_p}{L_1 L_2} + \left[\frac{\left(L_1 + \mu_p\right)\left(L_1 + \mu_f\right)}{L_1\left(L_1 - L_2\right)}\right]e^{L_1 t} - \left[\frac{\left(L_2 + \mu_p\right)\left(L_2 + \mu_f\right)}{L_2\left(L_1 - L_2\right)}\right]e^{L_2 t} \tag{12.35}$$

$$P_p(t) = \frac{\lambda_p \mu_f}{L_1 L_2} + \left[\frac{\lambda_p L_1 + \lambda_p \mu_f}{L_1\left(L_1 - L_2\right)}\right]e^{L_1 t} - \left[\frac{\left(\mu_f + L_2\right)\lambda_p}{L_2\left(L_1 - L_2\right)}\right]e^{L_2 t} \tag{12.36}$$

$$P_f(t) = \frac{\lambda_f \mu_p}{L_1 L_2} + \left[\frac{\lambda_f L_1 + \lambda_f \mu_p}{L_1\left(L_1 - L_2\right)}\right]e^{L_1 t} - \left[\frac{\left(\mu_p + L_2\right)\lambda_f}{L_2\left(L_1 - L_2\right)}\right]e^{L_2 t} \tag{12.37}$$

where

$$L_1 L_2 = \frac{-A \pm \left[A^2 - \left(\mu_p \mu_f + \lambda_f \mu_p + \lambda_p \mu_f \right) \right]^{1/2}}{2} \qquad (12.38)$$

$$A \equiv \mu_p + \mu_f + \lambda_p + \lambda_f \qquad (12.39)$$

$$L_1 + L_2 = -A \qquad (12.40)$$

$$L_1 L_2 = \mu_p \mu_f + \lambda_p \mu_f + \lambda_f \mu_p \qquad (12.41)$$

The probability of the system being down for preventive maintenance, corrective maintenance, and are given by Equation 12.36, Equation 12.37, and Equation 12.35, respectively.

As time t becomes large, we get the following steady state equations from Equation 12.35, Equation 12.36, and Equation 12.37, respectively:

$$P_0 = \frac{\mu_f \mu_p}{L_1 L_2} \qquad (12.42)$$

$$P_p = \frac{\lambda_p \mu_f}{L_1 L_2} \qquad (12.43)$$

$$P_f = \frac{\lambda_f \mu_p}{L_1 L_2} \qquad (12.44)$$

where P_0, P_p, and P_f are the steady-state probabilities of the system being in states 0, p, and f, respectively.

Example 12.4
Assume that in Equation 12.43, we have $\lambda_p = 0.0004$ per hour, $\mu_p = 0.0006$ per hour, $\lambda_f = 0.0001$ failures per hour, and $\mu_f = 0.0003$ repairs per hour. Calculate the steady-state probability that the system is down for preventive maintenance.

Inserting the above values into Equation 12.43 yields

$$P_p = \frac{(0.0004)(0.0003)}{(0.0006)(0.0003) + (0.0004)(0.0003) + (0.0001)(00006)}$$
$$= 0.3333$$

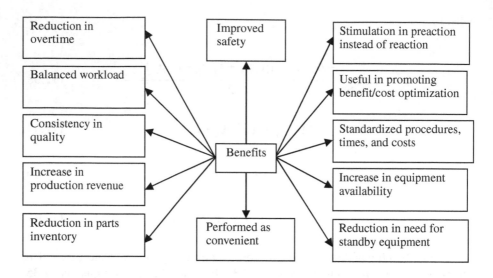

FIGURE 12.5 Important benefits of performing preventive maintenance.

Thus, there is an approximately 33% chance that the system will be down for preventive maintenance.

12.10 PREVENTIVE MAINTENANCE BENEFITS AND DRAWBACKS

There are many benefits of performing preventive maintenance. Most of the important benefits of performing preventive maintenance are shown in Figure 12.5 [13,14]. Some of the drawbacks of performing preventive maintenance are [13,14]:

- Exposing equipment to possible damage
- Increase in initial costs
- More frequent access to equipment
- Use of more components

12.11 PROBLEMS

1. Define the following terms:
   - Corrective maintenance
   - Preventive maintenance
2. Discuss the five types of corrective maintenance.
3. Discuss major corrective maintenance downtime components.
4. Discuss strategies for reducing the system-level corrective maintenance time.

5. Define the following corrective maintenance measures:
   - Mean corrective maintenance time.
   - Median active corrective maintenance time.
6. Assume that a system's mean time to failure is 3,000 hours and its mean corrective maintenance time, or mean time to repair, is 20 hours. Calculate the system steady-state unavailability if the system failure and repair times are exponentially distributed.
7. Discuss seven important elements of preventive maintenance.
8. Discuss steps for developing an effective preventive maintenance program in a short period.
9. Define the following preventive maintenance measures:
   - Mean preventive maintenance time.
   - Median preventive maintenance time.
10. What are the advantages of performing preventive maintenance?

REFERENCES

1. McKenna, T. and Oliverson, R., *Glossary of Reliability and Maintenance Terms*, Gulf Publishing, Houston, TX, 1997.
2. Dhillon, B.S., *Engineering Maintenance: A Modern Approach*, CRC Press, Boca Raton, FL, 2002.
3. Omdahl, R.P., *Reliability, Availability, and Maintainability (RAM) Dictionary*, ASQC Quality Press, Milwaukee, WI, 1988.
4. *Engineering Design Handbook: Maintenance Engineering Techniques*, AMCP 706-132, Department of Defense, Washington, DC, 1975.
5. Niebel, B.W., *Engineering Maintenance Management*, Marcel Dekker, New York, 1994.
6. *Maintenance of Supplies and Equipment*, MICOM 750-8, Department of Defense, Washington, DC, March 1972.
7. *Maintainability Engineering Handbook*, NAVORD OD 39223, Department of Defense, Washington, DC, June 1969.
8. Blanchard, B.S., Verma, D., and Peterson, E.L., *Maintainability*, John Wiley & Sons, New York, 1995.
9. *Engineering Design Handbook: Maintainability Engineering Theory and Practice*, AMCP-766-133, Department of Defense, Washington, DC, 1976.
10. Dhillon, B.S., *Design Reliability: Fundamentals and Applications*, CRC Press, Boca Raton, FL, 1999.
11. Shooman, M.L., *Probabilistic Reliability: An Engineering Approach*, McGraw-Hill, New York, 1968.
12. Levitt, J., *Maintenance Management*, Industrial Press, New York, 1997.
13. Levitt, J., Managing preventive maintenance, *Maintenance Technology*, February, 20–30, 1997.
14. Patton, J.D., *Preventive Maintenance*, Instrument Society of America, Research Triangle Park, NC, 1983.
15. Westerkemp, T.A., *Maintenance Manager's Standard Manual*, Prentice Hall, Paramus, NJ, 1997.

16. Wild, R., *Essentials of Production and Operations Management*, Holt, Rinehart, and Winston, London, 1985.
17. Dhillon, B.S., *Mechanical Reliability: Theory, Models, and Applications*, American Institute of Aeronautics and Astronautics, Washington, DC, 1988.
18. Dhillon, B.S., *Power System Reliability, Safety, and Management*, Ann Arbor Science Publishers, Ann Arbor, MI, 1983.
19. Dhillon, B.S., *Reliability Engineering in Systems Design and Operations*, Van Nostrand Reinhold Company, New York, 1983.

13 Reliability-Centered Maintenance

13.1 INTRODUCTION

Over the past few decades engineering maintenance has changed dramatically because of various factors including a rapid change in technology. Increasing emphasis is now being placed on reliability-centered maintenance (RCM) because an organization can benefit from RCM when its breakdowns account for more than 20 to 25% of the total maintenance workload [1].

RCM is a systematic methodology used to identify the preventive maintenance–related tasks necessary for realizing the inherent reliability of equipment at the lowest cost. The history of RCM can be traced back to 1968 when the U.S. Air Transport Association (ATA) prepared a document entitled "Maintenance Evaluation and Program Development" for use with the Boeing 747 aircraft [2]. Two years later, this document was revised to handle two other wide-body aircraft: DC-10 and L-1011 [3].

In 1974 the U.S. Department of Defense commissioned United Airlines to prepare a report on processes used by the civil aviation sector to develop maintenance programs for aircraft [4]. The resulting report was "Reliability Centered Maintenance." Many other publications on the subject have appeared since then. A detailed history of RCM is available in References 4 to 10.

13.2 RCM GOALS AND PRINCIPLES

Some of the important goals of RCM are [11]:

- To establish design-related priorities that can facilitate preventive maintenance in an effective manner
- To plan preventive maintenance tasks that can reinstate safety and reliability to their original levels in the event of system or equipment deterioration
- To gather the data necessary for design improvement of items with proven unsatisfactory original reliability
- To accomplish the above three goals with minimal total cost (i.e., including the cost of residual failures and the maintenance cost).

There are many principles of RCM, as shown in Figure 13.1 [5]. Principle I means that RCM is concerned more with maintaining system and equipment function than maintaining the functioning of individual components. Principle II means that there are three types of maintenance tasks: failure-finding, time-directed, and condition-directed. The failure-finding tasks are concerned with discovering hidden functions

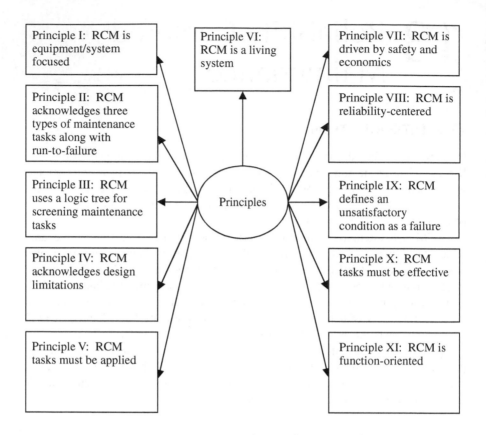

FIGURE 13.1 Principles of RCM.

that have failed to provide any indication of pending failures. The time-directed tasks are scheduled as considered appropriate. The condition-directed tasks are performed as the conditions indicate for their necessity. Run-to-failure is a conscious decision in RCM. Principle III is concerned with providing consistency in the maintenance of all types of equipment. Principle IV means that the goal of RCM is to maintain the inherent reliability of the product or equipment design. More specifically, maintenance at the best of times can only achieve and maintain the designed reliability. Principle V means that RCM tasks must lower the occurrence of malfunctions or ameliorate secondary damage resulting from failure. Principle VI means that RCM gathers information from the final results and feeds it back to enhance design and future maintenance. Principle VII means that safety is very important, so it must be ensured at any cost, and then cost effectiveness becomes the criterion. Principle VIII means that RCM is not overly concerned with simple failure rate but emphasizes the relationship between operating age and failures experienced in the field. More specifically, RCM treats failure statistics in an actuarial fashion. Principle IX means that a failure could be either a loss of function or a loss of acceptable quality. Principle X means that the maintenance tasks must be cost-effective and technically sound. Principle XI means that RCM is a pivotal factor in preserving equipment and system function, not just operability for its own sake.

13.3 RCM PROCESS-ASSOCIATED QUESTIONS AND RCM PROCESS

Many questions are associated with the RCM process. Any RCM process entails asking seven basic questions about the assets or system under review. [4,9,12]:

- What are the functions and associated standards of the asset performance in its current operating context?
- In what ways does it fail to meet its assigned functions?
- What are the specific causes for each functional failure?
- What are the specific effects of each malfunction or failure?
- In what specific way does each failure or malfunction matter?
- What possible actions can be taken to predict or prevent the occurrence of each failure?
- What measures can be exercised in the event of not finding a suitable proactive task?

The basic RCM process is made of seven steps, as shown in Figure 13.2 [13]. Step 1 calls for the identification of high-priority items with respect to maintenance

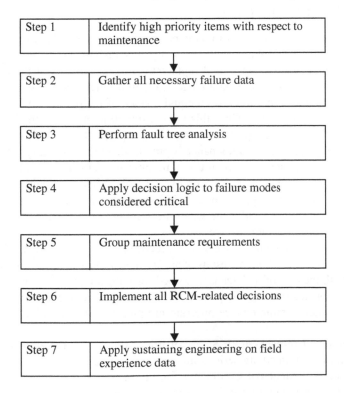

FIGURE 13.2 Basic RCM process steps.

by using methods such as failure mode, effects, and criticality analysis (FMECA) and fault tree analysis (FTA). Step 2 is concerned with collecting data on items such as part failure rates, operator error probabilities, and inspection efficiency from sources such as field experiences and generic failure databanks [14]. Step 3 involves calculating the probabilities of occurrence of fault events such as basic, intermediate, and top events, according to combinatorial properties of the logic elements in the fault tree. Step 4 is concerned with applying decision logic by asking standard assessment questions to the most desirable preventive maintenance task combinations and to each critical failure mode of each item important to maintenance. Step 5 involves classifying maintenance requirements into three groups: condition-monitoring maintenance requirements, hard-time maintenance requirements, and on-condition maintenance requirements. Step 6 is concerned with setting and enacting task frequencies and intervals as part of the overall maintenance strategy plan. Step 7 involves the reevaluation of all RCM-associated default decisions.

13.4 KEY RCM PROGRAM ELEMENTS

These are reactive maintenance, predictive testing and inspection, proactive maintenance, and preventive maintenance [9,15,16]. Each of these elements is described below.

13.4.1 REACTIVE MAINTENANCE

Other names used for this type of maintenance are breakdown, fix-when-fail, repair, and run-to-failure maintenance. When using this maintenance method, item and equipment repair, maintenance, and replacement occur only when the degradation in the condition of an item or equipment leads to a functional failure. This type of maintenance assumes that there is an equal chance of malfunction or failure in any component, part, or system. Thus, this very assumption precludes the identification of a certain class of repair parts as being more appropriate than others.

When only this type of maintenance is practiced, a high percentage of unplanned maintenance-related activities, poor use of maintenance effort, and high replacement part inventories are typical [15]. Moreover, a solely reactive maintenance–based program overlooks many opportunities to influence item and equipment survivability, and there is no ability to influence when the failures occur because of the absence of actions to control or prevent them.

However, reactive maintenance can be practiced effectively if it is performed as a conscious decision based on the results of an RCM analysis that compares risk and cost of failure with the cost of maintenance needed for mitigating that risk and failure cost.

Criteria for determining the priority for repairing or replacing the failed equipment in the reactive maintenance program are presented below [15].

- **Priority 1 (emergency):** Safety of life or property is threatened or there will be an immediate serious impact on mission.
- **Priority 2 (urgent):** Continuous facility operation is threatened or there will be an impending serious impact on mission.

- **Priority 3 (priority):** There will be degradation in quality of mission support or a significant and adverse effect on project.
- **Priority 4 (routine):** Redundancy is available or impact on mission will be insignificant.
- **Priority 5 (discretionary):** Impact on mission is negligible and resources are available.
- **Priority 6 (deferred):** Impact on mission is negligible and resources are not available.

13.4.2 PREDICTIVE TESTING AND INSPECTION

Occasional predictive testing and inspection (PTI) is also called predictive maintenance or condition monitoring and it uses basically nonintrusive testing methods, performance data, and visual inspection to assess item and equipment condition. PTI replaces arbitrarily timed maintenance tasks with maintenance that is carried out only when warranted by the condition of the item or equipment. The analysis of condition-monitoring data on a continuous basis permits planning and scheduling of maintenance or repairs in advance of functional and catastrophic failure.

PTI data are utilized in various ways to determine the condition of equipment and identify failure precursors. Six methods of analysis are shown in Figure 13.3 [15]. PTI must not be the sole type of maintenance practiced because it does not lend itself to all types of equipment or possible failure modes.

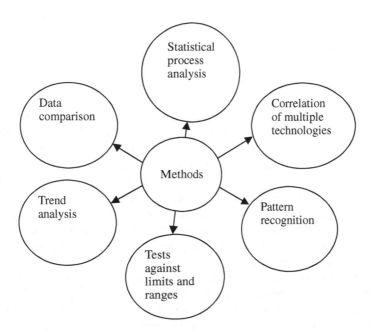

FIGURE 13.3 Methods used to analyze PTI data.

13.4.3 PROACTIVE MAINTENANCE

This type of maintenance improves maintenance through better design, installation, scheduling, workmanship, and maintenance procedures. The important characteristics of proactive maintenance are [15]:

- Use of feedback and communications in ensuring that any changes in design or procedures are rapidly made accessible to managers and designers
- Periodic evaluation of the technical matter and performance interval of maintenance tasks
- Use of a life-cycle view of maintenance and supporting functions
- Application of root-cause failure analysis and predictive analysis in maximizing maintenance effectiveness
- Assurance to a degree that nothing affecting maintenance occurs in isolation
- Adaptation of an ultimate goal of fixing the equipment or item forever
- Use of a continuous process of improvement
- Integration of functions that support maintenance into maintenance program planning
- Optimization and tailoring of appropriate methods and technologies to each specific application

A proactive maintenance program is the capstone of the RCM philosophy and it employs eight basic methods for extending equipment life: precision rebuild and installation, root-cause failure analysis, reliability engineering, age exploration, specifications for new and rebuilt equipment, failed-part analysis, recurrence control, and rebuild certification and verification. Each of these methods is described in detail in References 9 and 15.

13.4.4 PREVENTIVE MAINTENANCE

This type of maintenance is also known as time-driven or interval-based maintenance and is carried out without any regard to equipment condition. Preventive maintenance consists of regularly scheduled inspection, cleaning, adjustments, calibration, parts replacement, lubrication, and repair of components, equipment, and systems. Preventive maintenance schedules regular inspections and maintenance at predefined intervals in order to lower failures for susceptible items and equipment. It is important to note that, depending on the interval set, preventive maintenance can lead to a significant increase in inspections and routine maintenance. However, it can help to reduce the severity and frequency of unplanned equipment failures for parts with set, age-related wear patterns.

Traditional preventive maintenance is keyed to mean time between failures and failure rates and it assumes that such parameters can be evaluated statistically. Thus, one can replace an item due for failure prior to its failure. More specifically, failure rate and mean time between failures are frequently used in establishing the time interval for the performance of maintenance tasks. One major drawback of using these parameters to establish task periodicities is that failure rate data determines

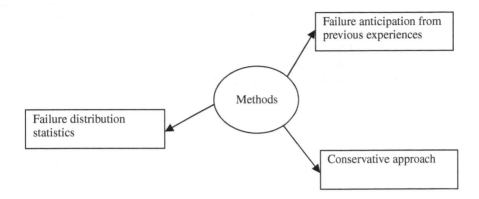

FIGURE 13.4 Useful methods for setting initial periodicities.

only the mean or average failure rate. Preventive maintenance can be ineffective or costly when it is the only type of maintenance practiced or used.

13.4.4.1 Determining Preventive Maintenance Tasks and Monitoring Periodicities

Although there are many different ways to determine the correct periodicity of preventive maintenance tasks, all these ways are incorrect without the proper knowledge of the in-service age-reliability characteristics of the item affected by the desired task. Normally, this type of information is not available but must be collected for new items or equipment. Past experiences indicate that PTI methods are quite effective in determining item condition versus age. When good information on the effect of age on equipment or item reliability is unavailable, the most effective approach is to monitor the equipment or item condition.

The main objective of monitoring an item's condition is to establish a trend for forecasting its future condition. For the purpose of setting initial periodicities, the three methods shown in Figure 13.4 are considered quite useful [15]. The failure anticipation from previous experiences approach is based on the reasoning that for some items failure history and personal experiences can provide an intuitive feel for when to expect a failure. The failure-distribution statistics approach is used to determine the basis for selecting periodicities and it requires full knowledge of item failure distribution and probability of failure. The conservative approach is commonly used in the industrial sector and is concerned with monitoring the item or equipment biweekly or monthly when good monitoring approaches and sufficient information are unavailable. Often, this leads to excessive monitoring.

13.5 RCM PROGRAM MEASURES

Many indicators have been developed to measure the effectiveness of RCM programs. Numerical indicators are considered to be the most effective because they are quantitative, precise, objective, and more easily trended than words and they consist of a benchmark and a descriptor [15]. A benchmark is a numerical expression of a set objective or goal. A descriptor is a word or a group of words that describe the function, the units, or the process under consideration for measurement.

This section presents a number of indicators considered useful for measuring the effectiveness of an RCM program, along with their suggested benchmark values. These benchmark values are the averages of data surveyed from approximately 50 major multinational corporations in the early 1990s [15].

13.5.1 INDEX 1

This index is used to calculate emergency percentage and is defined by

$$EMP = \frac{THWEJ}{THW} \tag{13.1}$$

where EMP is the emergency percentage, $THWEJ$ is the total hours worked on emergency jobs, and THW is the total hours worked.

The benchmark value for this index is 10% or less.

13.5.2 INDEX 2

This index is used to calculate maintenance overtime percentage and is expressed by

$$MOP = \frac{TOGP}{TRGP} \tag{13.2}$$

where MOP is the maintenance overtime percentage, $TOGP$ is the total number of maintenance overtime hours worked during a given period, and $TRGP$ is the total number of regular maintenance hours worked during a given period.

The benchmark value for this index is 5% or less.

13.5.3 INDEX 3

This index is used to calculate equipment availability (in percentage) and is expressed by

$$EQA = \frac{THEA}{THRP} \tag{13.3}$$

where EQA is the equipment availability, $THEA$ is the total number of hours each unit of equipment is available to run at capacity, and $THRP$ is the total number of hours during the reporting period.

The benchmark value for this index is 96%.

13.5.4 INDEX 4

This index is used to calculate the percentage of candidate equipment covered by PTI and is defined by

$$PCE = \frac{TEPP}{TECP} \tag{13.4}$$

where *PCE* is the percentage of candidate equipment covered by PTI, *TEPP* is the total number of equipment items in the PTI program, and *TECP* is the total number of equipment candidates for PTI.

The benchmark value for this indicator is 100%.

13.5.5 INDEX 5

This index is used to calculate the percentage of emergency work to PTI and preventive maintenance work and is expressed by

$$PEWP = \frac{TNEH}{TNPPH} \tag{13.5}$$

where *PEWP* is the percentage of emergency work to PTI and preventive maintenance work, *TNEH* is the total number of emergency work hours, and *TNPPH* is the total number of PTI and preventive maintenance work hours.

The benchmark value for this metric is 20% or less.

13.6 RCM BENEFITS AND CAUSES FOR RCM METHODOLOGY FAILURES

There are many benefits to using RCM. Some of the important ones are shown in Figure 13.5 [4,9,13]. Past experiences indicate that sometimes the application of RCM has also resulted in failure because of factors such as [4]:

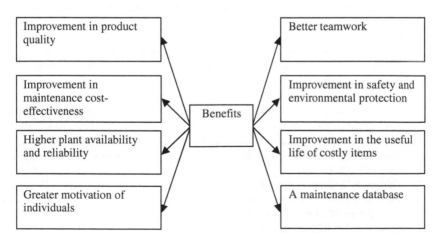

FIGURE 13.5 Important RCM benefits.

- Too much emphasis on failure data
- Superfluous or hurried RCM application
- Analysis performed at too low a level
- Only one person assigned to apply the RCM
- RCM applied by only the maintenance department
- RCM applied by manufacturers or vendors
- Use of computers to drive the process

13.7 PROBLEMS

1. Define the term *RCM*.
2. List at least four goals of RCM.
3. Discuss RCM principles.
4. Discuss the history of RCM.
5. Describe the RCM process.
6. What are the key RCM program elements? Discuss at least two of these elements in detail.
7. List at least five methods used to analyze predictive testing and inspection data.
8. Define the term *preventive maintenance*.
9. Define at least two indices used to measure RCM program effectiveness.
10. What are the causes for the failure of RCM?

REFERENCES

1. Picknell, J. and Steel, K.A., Using a CMMS to support RCM, *Maintenance Technology*, October, 110–117, 1997.
2. *Maintenance Evaluation and Program Development: 747 Maintenance Steering Group Handbook*, MSG1, Air Transport Association, Washington, DC, 1968.
3. *Airline/Manufacturer Maintenance Program Planning Document*, MSG2, Air Transport Association, Washington, DC, 1970.
4. Moubray, J., *Reliability Centered Maintenance*, Industrial Press, New York, 1992.
5. Nowlan, F.S. and Heap, H.F., *Reliability Centered Maintenance*, Dolby Access Press, San Francisco, 1978.
6. *Reliability Centered Maintenance Guide for Facilities and Collateral Equipment*, National Aeronautics and Space Administration (NASA), Washington, DC, 2000.
7. Smith, A.M., *Reliability Centered Maintenance*, McGraw-Hill, New York, 1993.
8. August, J., *Applied Reliability Centered Maintenance*, PennWell, Tulsa, OK, 1999.
9. Dhillon, B.S., *Engineering Maintenance: A Modern Approach*, CRC Press, Boca Raton, FL, 2002.
10. Dhillon, B.S., *Engineering Maintainability*, Gulf Publishing, Houston, TX, 1999.
11. *Guide to Reliability Centered Maintenance*, AMC Pamphlet No. 750-2, Department of the Army, Washington, DC, 1985.
12. Netherton, D., RCM tasks, *Maintenance Technology*, July/August 61–69, 1999.

13. Brauer, D.C. and Brauer, G.D., Reliability-centered maintenance, *IEEE Transactions on Reliability*, 36, 17–24, 1987.
14. Dhillon, B.S. and Viswanath, H.C., Bibliography of literature on failure data, *Microelectronics and Reliability*, 30, 723–750, 1990.
15. *Reliability Centered Maintenance Guide for Facilities and Collateral Equipment*, National Aeronautics and Space Administration (NASA), Washington, DC, 1996.
16. *Guidelines for the Naval Air Command*, NAVAIR 00-25-403, Department of Defense, Washington, DC, 1996.

14 Maintenance Management and Costing

14.1 INTRODUCTION

Just like in any other area of technology, management plays an important role in maintenance activity. Maintenance management is the function of providing policy guidance for all maintenance-related activities, in addition to exercising appropriate technical and management control of maintenance programs [1–3]. The effectiveness of maintenance management depends on many factors including the overall goal of the organization, the overall organizational set-up, training and skill of the maintenance management personnel, and training and skill of the personnel carrying out the maintenance activity.

A major proportion of the total equipment life cycle cost occurs during the maintenance phase. It has been estimated that the cost of maintaining equipment in the industrial sector varies from 2 to 20 times the acquisition cost [3]. Maintenance cost is the cost that includes lost opportunities in up time, yield, rate, and quality because of unsatisfactorily or nonoperating equipment, in addition to the cost associated with equipment-related degradation of the safety of the environment, people, and property [4].

14.2 MAINTENANCE MANAGEMENT PRINCIPLES

Six important principles of maintenance management are listed below [5,6]:

- The customer service relationship is the basis of an effective maintenance organization. Good maintenance service is very important for effectively maintaining facilities at an expected level. The team approach fostered by the organizational structure is quite important to consistent, active control of maintenance function.
- Maximum productivity occurs when each employee in an organization has a defined task to carry out in a definitive fashion and a definite time. This principle was formulated by Frederick Taylor in the late nineteenth century and it is still an important factor in management.
- Measurement comes before control. When a person is assigned a task to be performed using an effective method in a specified period of time, he or she becomes automatically aware of management expectations. Control begins when management personnel compare the results against set goals.

- Job control depends on definite, individual responsibility for each task during a work order's life span. A maintenance department's responsibility is to develop, implement, and provide appropriate operating support for the planning and scheduling of maintenance work. More specifically, it is the responsibility of management personnel to ensure effective and complete use of the system within their sphere of control.
- Schedule all control points effectively. Schedule appropriate control points at intervals so that all the problems are detected in time and the scheduled completion of the job is not delayed.
- The optimal size of a crew is the minimum number that can carry out a given task in an effective manner. Past experiences indicate that most tasks require just one person.

14.3 MAINTENANCE DEPARTMENT ORGANIZATION AND FUNCTIONS

Many factors play an instrumental role in determining the proper place of maintenance in the plant organization. Some of these factors are size, complexity, and the product produced. The four guidelines useful in planning a maintenance organization are shown in Figure 14.1 [7]. An important consideration in planning a maintenance organization is to decide whether to have a centralized or decentralized maintenance function. Some of the advantages of centralized maintenance are that it is more efficient than decentralized maintenance, needs fewer maintenance personnel, has more effective line supervision, and allows acquisition of more modern facilities and greater use of special equipment and specialized maintenance personnel [8]. The disadvantages of centralized maintenance include more difficult supervision because of remoteness of the maintenance site from the centralized headquarters, higher transportation cost because of remote maintenance work, more time spent getting to and from the work area [8].

Some of the advantages of decentralized maintenance are less travel time to and from maintenance jobs, usually closer supervision, and a spirit of cooperation

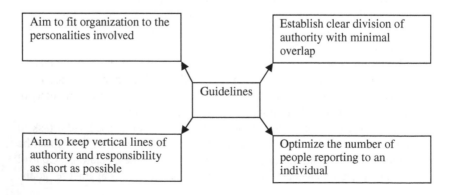

FIGURE 14.1 Useful guidelines for planning a maintenance organization.

between production and maintenance workers [8]. No one specific type of maintenance organization is useful for all types of enterprises [3].

A maintenance department performs a wide range of functions. Some of these functions are [7,9]:

- Preparing budgets with respect to material needs and maintenance personnel
- Keeping records on services, equipment, and so on
- Planning and repairing facilities to established standards
- Developing appropriate effective methods for monitoring maintenance staff activities
- Reviewing plans for establishing and constructing new facilities
- Preparing contract specifications
- Provide training to maintenance staff as the need arises
- Performing preventive maintenance
- Managing spare part inventory
- Developing appropriate methods for keeping all concerned people aware of maintenance activities
- Developing safety education programs for maintenance personnel
- Implementing appropriate methods for improving workplace safety
- Inspecting work carried out by contractors to ensure compliance with contractual requirements

14.4 EFFECTIVE MAINTENANCE MANAGEMENT ELEMENTS

The effectiveness of maintenance management depends on many elements. Some of these elements shown in Figure 14.2 are briefly discussed below [3,6]:

- **Maintenance policy:** This is very important for a clear understanding of the maintenance management program and for continuity of maintenance-related operations, regardless of the size of a maintenance organization. Normally, maintenance organizations have manuals that contain information on items such as policies, objectives, responsibilities, programs, authority, reporting requirements, performance measurements, and useful methods and techniques.
- **Work order system:** This is a useful tool to help management control costs and evaluate job performance. A work order authorizes and directs individuals to carry out an assigned task. Usually a work order contains information such as work description and associated reasons, requested and planned completion dates, planned start date, work category (i.e., repair, preventive maintenance, installation, etc.), items to be affected, appropriate approval signatures, and labor and material costs. A well-defined work order system should cover all types of

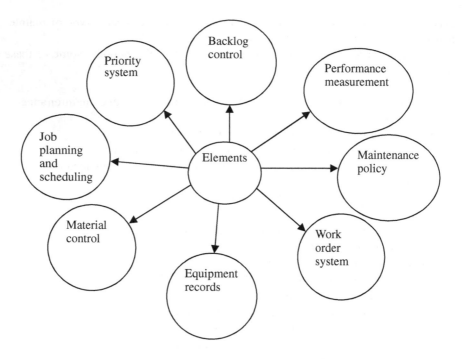

FIGURE 14.2 Elements of effective maintenance management.

maintenance jobs requested and completed, whether one-time or repetitive jobs.

- **Material control:** Effectiveness in material coordination is an important factor in efficient utilization of maintenance personnel. Material-related problems can result in delays in job completion, excess travel time, and so on. Some of the steps that can help reduce material-related problems are job planning, coordinating with stores, coordinating with purchasing, and reviewing the completed jobs. On average, material costs account for about 30 to 40% of total direct maintenance costs [6].

- **Job planning:** Job planning is an important element of effective maintenance management because prior to starting a maintenance job there could be a need to perform a number of tasks, for example, procurement of appropriate components, materials, and tools; securing safety permits; coordination with other departments; and coordination and delivery of parts, tools, and materials. Although the degree of planning needed may depend on the craft involved and methods used, past experiences indicate that usually 1 planner is required for every 20 craft persons. In most maintenance organizations 80 to 85% planning coverage can be attained.

- **Job scheduling:** Scheduling is as important as maintenance job planning and its effectiveness depends on the reliability of the planning function.

For large maintenance jobs to assure effective overall control, the use of methods such as the critical path method (CPM) and the program evaluation and review technique (PERT) must be considered.

- **Backlog control:** The amount of backlog within a maintenance organization plays an important role in the effectiveness of maintenance management. Its identification is very important for balancing workload and personnel needs. In addition, decisions on items such as subcontracting maintenance work, overtime, shop assignments, and hiring are basically based on backlog information. Usually, management uses various indexes in making decisions concerning backlog.
- **Equipment records:** These are another important factor that plays an important role in the efficiency of the maintenance organization. Generally, equipment records are classified under four categories: inventory, maintenance cost, files, and maintenance work performed. Equipment records are used in various areas including troubleshooting breakdowns, investigating incidents, procuring new equipment to determine operating performance trends, performing life cycle cost and design studies, conducting replacement and modification studies, and conducting reliability and maintainability studies.
- **Performance measurement:** Progressive maintenance organizations measure their performance on a regular basis through various means. Performance analyses play an important role in maintenance organization efficiency and are useful in revealing equipment downtime, peculiarities in operational behavior of the organization, and so on. Maintenance management makes use of various types of indexes to measure performance.
- **Priority system:** In a maintenance organization, the determination of job priority is absolutely essential since it is not possible to start every maintenance job the day it is requested. In assigning job priorities, progressive maintenance management carefully considers factors such as the type of maintenance required, the importance of the equipment or item, required due dates, and the length of time the job awaiting scheduling will take.

14.5 QUESTIONS FOR EVALUATING MAINTENANCE PROGRAM EFFECTIVENESS

The U.S. Energy Research and Development Administration conducted a study of maintenance management-related matters [6]. As the result of this study, it formulated ten questions for maintenance managers to use to evaluate their ongoing maintenance efforts, as presented in Table 14.1 [6]. If an unqualified "yes" is the answer to each of the ten questions, your ongoing maintenance effort is on sound footing to meet the objectives of organization. Otherwise, the maintenance program needs appropriate corrective measures.

TABLE 14.1
Questions for Maintenance Managers to Self-Evaluate Their Maintenance Efforts

No.	Question
1	Are you aware of the activities and facilities that consume most of the maintenance dollars?
2	Are you fully aware of the amount of time your foreman or foremen spend at the desk and at the job site?
3	Are you aware of whether proper safety practices are being followed effectively?
4	Are you fully aware of how your craft persons spend their time, that is, delays, travel, etc.?
5	Have you balanced your inventory of spare parts in regard to anticipated downtime losses versus carrying cost?
6	Do you have an appropriate base for performing productivity measurements, and is productivity improving?
7	Are your craft persons provided with the right quantity and quality of material when and where they need it?
8	Are you fully aware of whether your craft persons use the correct methods and tools to carry out their assigned tasks?
9	Can you compare the "should" with the "what" with respect to job costs?
10	Do you ensure the consideration of maintainability factors in the design of new or modified equipment and facilities?

14.6 MAINTENANCE COSTING REASONS AND MAINTENANCE COST INFLUENCES

There are many reasons for maintenance costing. Some of the main ones are to determine maintenance cost drivers, to improve productivity, to control costs, to prepare budgets, to compare competing approaches to maintenance, to provide input to equipment life cycle cost studies, to develop optimum preventive maintenance policies, to provide input to the design of new facility or equipment, to make decisions concerning equipment replacement, to compare maintenance cost effectiveness to industry averages, and to provide feedback to upper-level management [10]. Some of the major factors that influence maintenance costs are shown in Figure 14.3 [3,10].

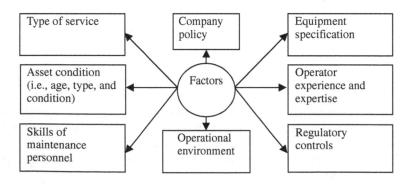

FIGURE 14.3 Some major factors that influence maintenance costs.

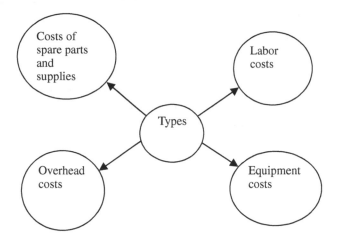

FIGURE 14.4 Types of cost data often collected by maintenance organizations.

14.7 COST DATA COLLECTION AND SOURCES

Maintenance costing requires various types of cost data. Management plays a key role in deciding on the type of data the maintenance organization should collect by considering its future application. Often, maintenance organizations collect four types of cost data, as shown in Figure 14.4 [9].

Equipment costs are used in making various types of decisions, and two important sources for obtaining these costs are the supplier's invoice and the purchase order.

Labor costs are usually obtained from the timesheet, and the overhead costs from the accounting department. Costs of spare parts and supplies are generally more difficult to obtain than the other three types of costs, but the work order is an important source for obtaining this cost data.

14.8 MAINTENANCE LABOR-COST ESTIMATION MODELS

Many mathematical models have been developed to estimate various types of maintenance labor costs. This section presents two such models.

14.8.1 CORRECTIVE-MAINTENANCE LABOR-COST ESTIMATION MODEL

This model estimates the annual labor cost of corrective maintenance when the item or equipment mean time to repair (MTTR) and mean time between failures (MTBF) are known. Thus, the annual corrective maintenance labor cost of a system is expressed by [5]

$$ALC_{cm} = \frac{(MTTR)(ASOH)(LCH)}{MTBF} \qquad (14.1)$$

where ALC_{cm} is the annual labor cost of corrective maintenance, $ASOH$ is the annual scheduled operating hours, and LCH is the corrective-maintenance labor cost per hour.

Example 14.1

A system's mean time between failures and mean time to repair are 500 hours and 10 hours, respectively. The system is scheduled to operate for 2,500 hours per year. Calculate the system annual corrective-maintenance labor cost if the hourly maintenance labor cost is $30.

Substituting the given data values into Equation 14.1 yields

$$ALC_{cm} = \frac{(10)(2,500)(30)}{500}$$
$$= \$1,500$$

Thus, the system annual corrective maintenance labor cost is $1,500.

14.8.2 TOTAL MAINTENANCE LABOR COST ESTIMATION MODEL

The total maintenance labor cost is expressed by [11]

$$TMLC = n\,(LRH)(TAH)(1+\alpha) \qquad\qquad (14.2)$$

where $TMLC$ is the total maintenance labor cost, n is the total number of employees, LRH is the labor rate per hour, TAH is the total number of annual hours, and α is the benefit ratio.

Example 14.2

Assume that for the maintenance department of a manufacturing company, we have the following data:

- $n = 10$ employees
- $\alpha = 0.3$
- LRH = $30
- TAH = 1,800 hours

Using Equation 14.2, calculate the company's total labor cost associated with the maintenance activity.

By substituting the specified data values into Equation 14.2, we get

$$TMLC = (10)\,(30)\,(1,800)\,(1+0.3)$$

$$= \$702,000$$

Thus, the company's total labor cost associated with maintenance activity is $702,000.

14.9 MAINTENANCE COST ESTIMATION MODELS FOR SPECIFIC EQUIPMENT OR FACILITY

Many mathematical models for estimating maintenance or related cost have been developed [11]. This section presents three such models.

14.9.1 MODEL 1

This model estimates the maintenance cost of an avionics computer. The avionics computer's maintenance cost is expressed by [12]

$$TMC_{ac} = \frac{(AMC_u)\,m}{1,000}$$ (14.3)

where TMC_{ac} is the total maintenance cost of the avionics computer, m is the number of years in operation, and AMC_u is the yearly maintenance cost per unit expressed in 1974 dollars ($\times 10^3$).

The natural logarithm of AMC_u is expressed by

$$\ln AMC_u = \theta_1 + \theta_2 \ln UC - \theta_3 \ln MTBF_{ac}$$ (14.4)

where $\theta_1 = 6.944$, $\theta_2 = 0.296$, $\theta_3 = -0.63$, UC is the unit cost expressed in 1974 dollars ($\times 10^3$), and $MTBF_{ac}$ is the avionics computer mean time between failures expressed in hours.

14.9.2 MODEL 2

This model estimates the maintenance cost of a fire control radar. The maintenance cost is expressed by [13]

$$FCRMC = \frac{(AFH)(m)(MCFH)}{1,000}$$ (14.5)

where $FCRMC$ is the fire control radar maintenance cost, AFH is the annual flying hours, and $MCFH$ is the maintenance cost per flying hour per unit expressed in 1974 dollars ($\times 10^3$).

The natural logarithm of MCFH is expressed by

$$\ln MCFH = \alpha_1 + \alpha_2 \ln P_p$$ (14.6)

where $\alpha_1 = -2.086$, $\alpha_2 = 0.611$, and P_p is the peak power in kilowatts.

14.9.3 MODEL 3

In a manufacturing company, often the cost of production facility downtime is also factored into the maintenance cost. This model estimates the cost of production facility downtime. The production facility downtime cost is expressed by [10]

$$DC_{pf} = RC + RPRC + RLC + LPC + IOS + TAIC$$ (14.7)

where *RC* is the rental cost of replacement unit (if any); *RPRC* is the ruined product replacement cost; *RLC* is the revenue loss cost, less recoverable costs like materials; *LPC* is the late penalty cost, *IOS* is the idle operator salary, and *TAIC* is the tangible and intangible costs associated with factors such as customer dissatisfaction and loss of good will.

14.10 PROBLEMS

1. Discuss six important principles of maintenance management.
2. List at least ten important functions of a maintenance engineering department.
3. Discuss the six most important elements of maintenance management.
4. What are the ten questions for managers to evaluate their maintenance program effectiveness?
5. What are the important reasons for maintenance costing?
6. List at least seven major factors that influence maintenance cost.
7. Discuss four types of cost data often collected by maintenance organizations.
8. Discuss advantages and disadvantages of centralized maintenance.
9. Discuss useful guidelines for planning a maintenance organization.
10. Assume that an equipment's mean time between failures and mean time to repair are 400 hours and 5 hours, respectively. The equipment is scheduled to operate for 3,000 hours annually. Calculate the equipment annual corrective maintenance labor cost if the hourly maintenance labor cost is $25.

REFERENCES

1. *Engineering Design Handbook: Maintenance Engineering Techniques*, AMCP 706–132, Department of the Army, Washington, DC, 1975.
2. *Policies Governing Maintenance Engineering Within the Department of Defense*, DOD Inst. 4151.12, Washington, DC, 1968.
3. Dhillon, B.S., *Engineering Maintenance: A Modern Approach*, CRC Press, Boca Raton, FL, 2002.
4. McKenna, T. and Oliverson, R., *Glossary of Reliability and Maintenance Terms*, Gulf Publishing, Houston, TX, 1997.
5. Westerkamp, T.A., *Maintenance Manager's Standard Manual*, Prentice Hall, Paramus, NJ, 1997.
6. *Maintenance Manager's Guide*, ERHQ-0004, Energy Research and Development Administration, Washington, DC, 1976.
7. Higgins, L.R., *Maintenance Engineering Handbook*, McGraw-Hill, New York, 1988.
8. Niebel, B.W., *Engineering Maintenance Management*, Marcel Dekker, New York, 1994.
9. Jordan, J.K., *Maintenance Management*, American Water Works Association, Denver, CO, 1990.
10. Levitt, J., *The Handbook of Maintenance Management*, Industrial Press, New York, 1997.

11. Dhillon, B.S., *Life Cycle Costing: Techniques, Models, and Applications*, Gordon and Breach Science Publishers, New York, 1988.
12. Earles, M.E., *Factors, Formulas, and Structures for Life Cycle Costing*, Eddin-Earles Publishers, Concord, MA, 1978.
13. *Cost Analysis of Avionics Equipment*, Vol. 1, Prepared by the Air Force Systems Command, Wright-Patterson Air Force Base, OH, 1974 (NTIS Report No. AD 781132, available from the National Technical Information Service (NTIS), Springfield, VA).

15 Human Error in Engineering Maintenance

15.1 INTRODUCTION

Human interactions are an important factor during the design, installation, production, and maintenance phases of a product. Although the degree of these interactions may vary from one product to another and from one product phase to another, they are subject to deterioration because of human error.

While human error has existed since the beginning of humankind, only in the past 50 years has it been the subject of scientific inquiry. In regard to engineering products, human error is the failure to carry out a specified task (or the performance of a forbidden action) that could result in disruption of scheduled operations or damage to property and equipment [1–3].

Human errors may be grouped under six distinct categories: operating errors, assembly errors, design errors, inspection errors, installation errors, and maintenance errors [1–5]. Maintenance error is the result of the wrong preventive or repair actions. Usually, the probability of occurrence of human error increases along with the increase in maintenance frequency as the product or equipment ages [1].

15.2 MAINTENANCE ERROR-RELATED FACTS AND FIGURES

Some maintenance error-related facts and figures are as follows:

- In 1979, 272 people died in an aircraft accident because of improper maintenance procedures [6].
- From 1982 to 1991, a study of safety issues concerning onboard fatalities of a worldwide jet fleet revealed that maintenance and inspection was the second most important safety concern for 1,481 onboard fatalities [7–8].
- A study of 213 maintenance problem reports revealed that about 25% were due to human error [2,9].
- A study of maintenance operations among commercial airlines reported that about 40 to 50% of the time, the parts that were removed for repair were not defective at all [6].
- A study of maintenance errors occurring in missile operations reported many causes for their occurrence [2,6]: wrong installation (28%), dials

and controls (missed or misread) (38%), inaccessibility (3%), loose nuts or fittings (14%), and miscellaneous (17%) [2,6].

- A study of maintenance tasks such as aligning, removing, and adjusting reported an average human reliability of 0.9871 [10]. This means one should expect around 13 errors in every 1,000 such maintenance tasks [6].
- In 1983, a passenger aircraft departing Miami, Florida, lost oil pressure in all three of its engines as a result of chip detector O-rings that were missing because of poor inspection and supply procedures followed by maintenance personnel [11].

15.3 REASONS FOR HUMAN ERROR IN MAINTENANCE

Human error in maintenance occurs for many reasons. Some of the important ones are shown in Figure 15.1 [2,6]. In particular, with regard to training and experience of maintenance technicians, a study reported that those who ranked highest possessed characteristics such as [6,10]:

- Higher morale
- More work experience
- Fewer reports of fatigue
- Greater satisfaction with the work group
- Higher aptitude

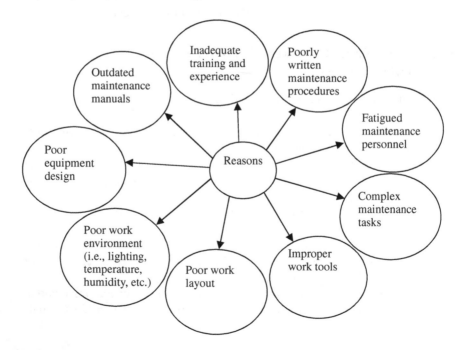

FIGURE 15.1 Some of the important reasons for the occurrence of human error in maintenance.

In addition, correlation analysis has reported positive correlations between task performance and factors such as amount of time in career field, ability to handle responsibility, morale, and years of experience. Negative correlations were found between task performance and fatigue symptoms and anxiety level.

15.4 MAJOR HUMAN FAILURES IN MAINTENANCE AND CLASSIFICATIONS OF MAINTENANCE ERRORS AND THEIR OCCURRENCE FREQUENCY

Many studies of human factors in airline maintenance have been conducted. One of these studies has identified the following eight major human failures in maintenance of aircraft over 5,700 kg [8]:

* Installation of incorrect parts
* Poor lubrication
* Discrepancies in electrical wiring
* Fitting of wrong parts
* Leaving loose objects in the aircraft
* Unsecured oil or fuel caps and refuel panels
* Failure to remove landing gear ground lock pins prior to departure
* Unsecured access panels, fairings, and cowlings

In the early 1990s Boeing conducted a study of 86 incident reports with respect to maintenance error. It classified human errors in maintenance into 31 distinct categories. These categories, along with their corresponding occurrence frequencies in parentheses are as follows [12]:

* System operated in unsafe conditions (16)
* System not made safe (10)
* Equipment failure (10)
* Towing event (10)
* Falls and spontaneous actions (6)
* Degradation not found (6)
* Person entered dangerous area (5)
* Incomplete installation (5)
* Work not documented (5)
* Person did not obtain or use appropriate equipment (4)
* Person contacted hazard (4)
* Unserviceable equipment used (4)
* System not reactivated or deactivated (4)
* Verbal warning not given (3)
* Safety lock or warning removed (2)
* Pin or tie left in place (2)
* Not properly tested (2)
* Vehicle or equipment contacted aircraft (2)
* Warning sign or tag not used (2)
* Vehicle driving (not towing) (2)

- Wrong fluid type (1)
- Access panel not closed (1)
- Panel installed incorrectly (1)
- Material left in aircraft or engine (1)
- Incorrect orientation (1)
- Equipment not installed (1)
- Contamination of open system (1)
- Wrong equipment or part installed (1)
- Person unable to access part in stores (1)
- Required servicing not performed (1)
- Miscellaneous (6)

15.5 HUMAN ERROR IN MAINTENANCE PREDICTION MODELS

Many mathematical models can be used to predict the occurrence of human error in maintenance [2]. This section presents two such models.

15.5.1 MODEL 1

This mathematical model can be used to predict the probability of a maintenance person making an error. The model state–space diagram is shown in Figure 15.2. The numerals in boxes denote system states. The following assumptions are associated with this model [13]:

- The maintenance person is performing a time-continuous task.
- The rate of errors made by the maintenance person is constant.
- The errors occur independently.

The following symbols are associated with Figure 15.2:

- λ_m is the constant maintenance error rate (i.e., the rate of errors made by the maintenance person).
- $P_0(t)$ is the probability that the maintenance person is performing his or her task correctly at time t.
- $P_1(t)$ is the probability that the maintenance person has committed an error at time t.

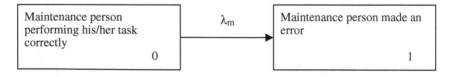

FIGURE 15.2 State–space diagram of a maintenance person performing a time-continuous task.

Using the Markov Method, we write down the following equations for Figure 15.2 [13]:

$$\frac{dP_0(t)}{dt} + \lambda_m P_0(t) = 0 \tag{15.1}$$

$$\frac{dP_1(t)}{dt} - \lambda_m P_0(t) = 0 \tag{15.2}$$

At time $t = 0$, $P_0(0) = 1$ and $P_1(0) = 0$.
Solving Equation 15.1 and Equation 15.2, using Laplace transforms, we get

$$P_0(t) = e^{-\lambda_m t} \tag{15.3}$$

$$P_1(t) = 1 - e^{-\lambda_m t} \tag{15.4}$$

The maintenance person's reliability is given by

$$R_m(t) = P_0(t) = e^{-\lambda_m t} \tag{15.5}$$

where $R_m(t)$ is the maintenance person's reliability at time t.
Mean time to maintenance error (MTTME) is given by [13]

$$\begin{aligned} MTTME &= \int_0^\infty R_m(t)\,dt \\ &= \int_0^\infty e^{-\lambda_m t}\,dt \\ &= \frac{1}{\lambda_m} \end{aligned} \tag{15.6}$$

Example 15.1
A maintenance person is performing some time-continuous tasks and his or her error rate is 0.008 errors per hour. Calculate his or her reliability during a 7-hour mission.
Substituting the given data values into Equation 15.5 yields

$$\begin{aligned} R_m(7) &= e^{-(0.008)(7)} \\ &= 0.9455 \end{aligned}$$

Thus, the reliability of the maintenance person during the 7-hour mission is 0.9455.

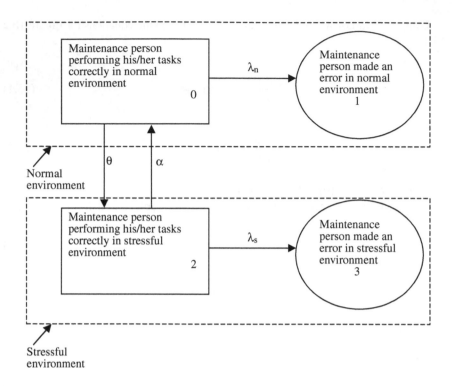

FIGURE 15.3 State–space diagram for the maintenance person performing tasks in a fluctuating environment.

15.5.2 Model 2

This mathematical model can be used to predict the probability of the maintenance person making an error in a fluctuating environment (i.e., normal or stressful). The model state–space diagram is shown in Figure 15.3 [2]. The numerals in boxes and circles denote system states. The following assumptions are associated with this model [2,13,14]:

- The maintenance person is performing time-continuous tasks in fluctuating environments.
- The rates of errors made by the maintenance person in fluctuating environment are different and constant.
- The errors occur independently.

The following symbols are associated with Figure 15.3:

- $P_i(t)$ is the probability of the maintenance person being in state i at time t for $i = 0$ (maintenance person performing tasks correctly in a normal environment), $i = 1$ (maintenance person made an error in a normal environment), $i = 2$ (maintenance person performing tasks correctly in a stressful environment), and $i = 3$ (maintenance person made an error in a stressful environment).

- λ_n is the constant error rate of the maintenance person when working in a normal environment.
- λ_s is the constant error rate of the maintenance person when working in a stressful environment.
- α is the constant transition rate from a stressful environment to a normal environment.
- θ is the constant transition rate from a normal environment to a stressful environment.

Using the Markov method, we write down the following equations for Figure 15.3 [13,14]:

$$\frac{dP_0(t)}{dt}+(\lambda_n+\theta)P_0(t)=\alpha P_2(t) \tag{15.7}$$

$$\frac{dP_1(t)}{dt}=\lambda_n P_0(t) \tag{15.8}$$

$$\frac{dP_2(t)}{dt}+(\lambda_n+\alpha)P_2(t)=\theta P_0(t) \tag{15.9}$$

$$\frac{dP_3(t)}{dt}=\lambda_s P_2(t) \tag{15.10}$$

At time $t = 0$, $P_0(0) = 1$ and $P_1(0) = P_2(0) = P_3(0) = 0$.

Solving Equation 15.7 to Equation 15.10, using Laplace transforms, results in the following state probability equations:

$$P_0(t)=(y_2-y_1)^{-1}\left[\left(y_2+\lambda_s+\alpha\right)e^{y_2 t}-\left(y_1+\lambda_s+\alpha\right)e^{y_1 t}\right] \tag{15.11}$$

where

$$y_1=\frac{-b_1+\left(b_1^2-4b_2\right)^{1/2}}{2} \tag{15.12}$$

$$y_2=\frac{-b_1-\left(b_1^2-4b_2\right)^{1/2}}{2} \tag{15.13}$$

$$b_1=\lambda_n+\lambda_s+\theta+\alpha \tag{15.14}$$

$$b_2=\lambda_n\left(\lambda_s+\alpha\right)+\theta\lambda_s \tag{15.15}$$

$$P_1(t)=b_4+b_5\ e^{y_2 t}-b_6\ e^{y_1 t} \tag{15.16}$$

where

$$b_3 = \frac{1}{y_2 - y_1} \tag{15.17}$$

$$b_4 = \lambda_n \left(\lambda_s + \alpha\right) / y_1 y_2 \tag{15.18}$$

$$b_5 = b_3 \left(\lambda_n + b_4 y_1\right) \tag{15.19}$$

$$b_6 = b_3 \left(\lambda_n + b_4 y_2\right) \tag{15.20}$$

$$P_2(t) = \theta b_3 \left(e^{y_2 t} - e^{y_1 t}\right) \tag{15.21}$$

$$P_3(t) = b_7 \left[\left(1 + b_3\right)\left(y_1 e^{y_2 t} - y_2 e^{y_1 t}\right)\right] \tag{15.22}$$

where

$$b_7 = \lambda_s \theta / y_1 y_2 \tag{15.23}$$

The maintenance person's reliability is given by

$$R_{mp}(t) = P_0(t) + P_2(t) \tag{15.24}$$

The mean time to maintenance error (MTTME) is given by [13]

$$\begin{aligned} MTTME &= \int_0^\infty R_{mp}(t)\,dt \\ &= \left(\lambda_s + \theta + \alpha\right) / b_2 \end{aligned} \tag{15.25}$$

Example 15.2
Assume that a maintenance person is performing maintenance tasks in a fluctuating environment (i.e., normal and stressful). The constant error rates of the maintenance person under normal and stressful conditions are 0.002 errors per hour and 0.007 errors per hour, respectively. The value of the transition from a normal environment to a stressful environment is 0.04 per hour, and conversely, 0.01 per hour.
    Calculate the value of the mean time to maintenance error.

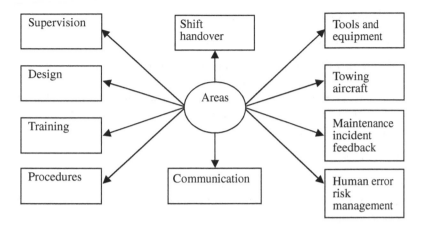

FIGURE 15.4 Areas covered by guidelines for reducing human error in airline maintenance.

Inserting the specified data values into Equation 15.25 yields

$$MTTME = \frac{\lambda_s + \theta + \alpha}{\lambda_n(\lambda_s + \alpha) + \theta\lambda_s}$$

$$= \frac{0.007 + 0.04 + 0.01}{(0.002)(0.007 + 0.01) + (0.04)(0.007)}$$

$$= 181.53 \ hours$$

Thus, the mean time to maintenance error is 181.53 hours.

15.6 USEFUL GUIDELINES FOR REDUCING MAINTENANCE ERRORS

Various guidelines have been developed for reducing the occurrence of human error in maintenance. This section presents guidelines developed to reduce the occurrence of human error in airline maintenance. Many of these guidelines can also be used in other types of maintenance. The guidelines cover 10 specific areas as shown in Figure 15.4 [8]. Guidelines for each of these areas are presented below.

15.6.1 DESIGN

Two important guidelines pertaining to design are:

- Actively seek information on human errors occurring during the maintenance phase to provide input in the design phase.
- Ensure that manufacturers give adequate attention to maintenance-related human factors during the design process.

15.6.2 Tools and Equipment

Two useful guidelines concerning tools and equipment are:

- Ensure that the storage of lock-out devices becomes immediately clear if they are left in place inadvertently.
- Review systems by which items such as stands and lighting systems are maintained for the removal of unserviceable equipment from service and repairing them rapidly.

15.6.3 Towing Aircraft

A useful guideline with respect to towing aircraft or other equipment is to review, in an effective manner, the procedures and equipment used for towing to and from maintenance facilities.

15.6.4 Procedures

Two useful guidelines concerning procedures are:

- Review work practices periodically with respect to their significant departure from formal procedures.
- Ensure the following of standard work practices across all maintenance operations.

15.6.5 Training

Two important guidelines pertaining to training are:

- Periodically provide maintenance personnel with refresher training courses that emphasize company procedures.
- Seriously consider introducing crew resource management for maintenance personnel and associated personnel, that is, individuals who interact with the maintenance professionals.

15.6.6 Human Error Risk Management

Some useful guidelines pertaining to human error risk management are:

- Avoid performing simultaneously the same maintenance task on redundant units.
- Seriously consider the need to disturb normally operating items for performing nonessential periodic maintenance inspections because the disturbance may lead to a maintenance error.

15.6.7 Maintenance Incident Feedback

Two useful guidelines related to this area are:

- Ensure that people involved with training receive regular feedback on recurring maintenance incidents so that appropriate corrective measures for these problems are targeted effectively.

• Ensure that management receives regular feedback on maintenance incidents with particular consideration to the underlying conditions that cause such incidents.

15.6.8 SUPERVISION

A useful guideline in this area is to recognize that oversights by supervisors and management need to be strengthened, particularly in the last few hours of each shift as the occurrence of errors becomes more likely.

15.6.9 COMMUNICATION

An important guideline concerning communication is to ensure that appropriate systems are in place for disseminating essential information to all maintenance personnel so that repeated errors or changing procedures are considered carefully.

15.6.10 SHIFT HANDOVER

A useful guideline in this area is to ensure the adequacy of practices associated with shift handover by considering communication and documentation so that incomplete tasks are transferred correctly and effectively across all shifts.

15.7 PROBLEMS

1. Define human error.
2. What are the main reasons for the occurrence of human error in maintenance?
3. List at least five maintenance error–related facts and figures.
4. List major human failures in the maintenance of aircraft over 5,700 kg.
5. Discuss the classifications of maintenance errors.
6. Write an essay on human error in engineering maintenance.
7. List at least 10 useful guidelines for reducing human errors in maintenance.
8. A person is performing some maintenance tasks and his or her error rate is 0.0024 errors per hour. Calculate the person's reliability during an 8-hour time period.
9. A maintenance person is performing maintenance tasks in a fluctuating environment (i.e., normal and stressful). The constant error rates of the maintenance person under normal and stressful conditions are 0.004 errors per hour and 0.009 errors per hour, respectively. The value of the transition rate from normal to stressful environments is 0.05 per hour, and conversely, 0.02 per hour. Calculate the value of the mean time to maintenance error.
10. From problem 9, calculate the maintenance person's reliability during an 8-hour mission.

REFERENCES

1. Meister, D., *Human Factors: Theory and Practice*, John Wiley & Sons, New York, 1976.
2. Dhillon, B.S., *Human Reliability: With Human Factors*, Pergamon Press, New York, 1986.
3. Meister, D., The problem of human-initiated failures, *Proceedings of the 8th National Symposium on Reliability and Quality Control*, 1962, pp. 234–239.
4. Hagen, E.W., Ed., Human reliability analysis, *Nuclear Safety*, 17, 315–326, 1976.
5. Dhillon, B.S., *Engineering Maintenance: A Modern Approach*, CRC Press, Boca Raton, FL, 2002.
6. Christensen, J.M. and Howard, J.M., Field experience in maintenance, in *Human Detection and Diagnosis of System Failures*, Rasmussen, J. and Rouse, W.B., Eds., Plenum Press, New York, 1981, pp. 111–113.
7. Russell, P.D., Management strategies for accident prevention, *Air Asia*, 6, 31–41, 1991.
8. *Human Factors in Airline Maintenance: A Study of Incident Reports*, Bureau of Air Safety Investigation, Department of Transport and Regional Development, Canberra, Australia, 1997.
9. Robinson, J.E., Deutsch, W.E., and Rogers, J.G., The field maintenance interface between human engineering and maintainability engineering, *Human Factors*, 12, 253–259, 1970.
10. Sauer, D., Campbell, W.B., Potter, N.R., and Askern, W.B., *Relationships Between Human Resource Factors and Performance on Nuclear Missile Handling Tasks*, Report No. AFHRL-TR-76-85/AFWL-TR-76-301, Air Force Human Resources Laboratory/Air Force Weapons Laboratory, Wright-Patterson Air Force Base, OH, 1976.
11. Tripp, E.G., Human factors in maintenance, *Aviation Week's Business & Commercial Aviaton B/CA*, July, 1–10, 1999.
12. *Maintenance Error Decision Aid (MEDA)*, developed by Boeing Commercial Airplane Group, Seattle, WA, 1994.
13. Dhillon, B.S., *Design Reliability: Fundamentals and Applications*, CRC Press, Boca Raton, FL, 1999.
14. Dhillon, B.S., Stochastic models for predicting human reliability, *Microelectronics and Reliability*, 21, 491–496, 1982.

16 Software Maintenance, Robotic Maintenance, and Medical Equipment Maintenance

16.1 INTRODUCTION

The field of maintenance has developed to a level where it has started to branch out into many specialized areas including software maintenance, robotic maintenance, and medical equipment maintenance.

Software maintenance is the process of making changes to the software system or component subsequent to delivery to improve performance or other attributes, rectify faults, or adapt to a change in the use environment [1,2]. In the early years of computing, software maintenance was only a small element of the overall software life cycle, but in recent years it has become a major factor. For example, in 1955 the proportion of time spent on maintenance activities was about 23%; in 1970 it increased to about 36%, and the prediction for 1985 was 58% [3,4]. In the mid-1980s the United States spent about $30 billion annually on software maintenance [5].

Although robots are generally reliable, sometimes they do fail and require maintenance, just as is the case for any other sophisticated machines. Thus, the users of robots must devise effective maintenance programs; otherwise their unscheduled downtime may increase to a point at which it defeats the purpose of robot applications. Moreover, careful consideration to maintenance should be given not only during the operational phase of robots but also during their design phase because various decisions regarding maintenance are made during this phase [6].

Medical equipment maintenance is also very important. More specifically, it is an important factor in providing effective health care. For example, poor maintenance of medical equipment can lead to high health-care costs and patient deaths [7].

16.2 SOFTWARE MAINTENANCE FACTS AND FIGURES

Some of the facts and figures associated with software maintenance are as follows:

- Software maintenance activities account for about 70% of the overall software cost [8].
- Over 80% of the life of a software product is spent in maintenance [9].
- Modifications and extensions requested by users account for over two-fifths of software maintenance activities [8].

- A study performed by Hewlett-Packard reported that about 60 to 80% of its software research and development staff members are involved in maintenance of existing software [10].
- The maintenance of existing software can consume over 60% of all development-related efforts [8].
- It is estimated that for all software systems combined, the maintenance component of the overall effort is increasing approximately 3% annually [11].
- A study performed by the Boeing Company reported that annually, on average, 15% of the lines of source code in simple programs are changed, 5% are changed in medium programs, and 1% are changed in difficult programs [11].

16.3 SOFTWARE MAINTENANCE TYPES

Software maintenance may be broken down under four classifications: perfective maintenance, corrective maintenance, preventive maintenance, and adaptive maintenance [12]. Perfective maintenance is concerned with adding capabilities, modifying existing functions, and making general enhancements. Corrective maintenance incorporates diagnosis and rectification of errors. Preventive maintenance is concerned with modifying software to enhance potential reliability and maintainability or provide an improved basis for future enhancements. Adaptive maintenance is concerned with modifying software to effectively interface with a changing environment (i.e., both hardware and software).

A survey of 487 software organizations reported the percentage distribution of the above types of maintenance as 50%, 21%, 4%, and 25%, respectively [13].

16.4 SOFTWARE MAINTENANCE TOOLS AND GUIDELINES FOR REDUCING SOFTWARE MAINTENANCE

Many methods have been developed that directly or indirectly concern software maintenance [14]. Two of these methods are presented below.

16.4.1 IMPACT ANALYSIS

Software maintenance depends on and begins with user needs. Past experiences indicate that a need translating into a seemly minor change is often more extensive and costly than anticipated. Under such circumstance, impact analysis is a useful tool to use to determine the risks related to the proposed change, including the estimation of effects on factors such as schedule, effort, and resources. Reference 16 presents various ways to measure the impact of a given change.

16.4.2 SOFTWARE CONFIGURATION MANAGEMENT

During software maintenance, keeping track of changes and their effects on other system components is a challenging task, and software configuration management is an effective tool to meet this challenge. Software configuration management is a set of tracking and control activities that start at the beginning of a software development project and terminate at the software retirement.

Configuration management is practiced by establishing a configuration control board because many software maintenance–associated changes are requested by users to rectify failures or make enhancements. The board oversees the change process, and its members are users, customers, and developers. Each problem is handled the following ways [12]:

- Software users, customers, or developers find a problem and use a formal change control form to record all its associated symptoms and information.
- The proposed change is formally reported to the configuration control board.
- The board members discuss the proposed change.
- The board makes a decision on the proposed change, prioritizes it, and assigns individuals to make the change.
- These individuals identify the problem source, highlight the changes required, and then test and implement the changes.
- The designated people work along with the software program librarian to track and control the change in the operational system and update related documentation.
- The designated individual files the report describing the changes made.

This method is described in detail in Reference 16.

Some important guidelines for reducing software maintenance are as follows [17–23]:

- As much as possible, use portable languages, tools, and operating systems.
- Establish effective communication among maintenance programmers.
- Identify all possible software enhancements and design the software so that it can easily incorporate such enhancements.
- Use standard methodologies.
- Employ preventive maintenance approaches such as using limits for tables that are reasonably greater than can possibly be needed.
- Divide the functions into two groups: inherently more stable and most likely to be changed.
- Store constants in tables rather than scattering them throughout the software program.
- Carefully consider human factors in areas that are the sources of frequent changes or modifications such as screen layouts.
- Introduce structured maintenance that uses methods for documenting existing systems, and incorporates guidelines for reading programs, and so on.

16.5 SOFTWARE MAINTENANCE COST ESTIMATION MODELS

Many mathematical models have been developed to estimate various types of software maintenance cost. This section presents two such models that can directly or indirectly be used to estimate software maintenance cost.

16.5.1 MODEL 1

In this case the software maintenance cost is expressed by [24,25]:

$$SMC = \left[3\left(C_m\right)k\right]/\alpha \qquad (16.1)$$

where k is the total number of instructions to be changed per month; C_m is the cost per person-month; α is the difficulty constant and its specified values are 500, 250, and 100 for easy, medium, and hard programs, respectively; and SMC is the software maintenance cost.

16.5.2 MODEL 2

This is a quite useful model to demonstrate how maintenance cost can build up alarmingly fast. The model is subject to the following two assumptions [26,27]:

- The programming work force is constant and is normalized to be unity.
- After the completion of the project, a (normalized) maintenance force, n, is assigned to perform maintenance activities. Consequently, a (normalized) work force, k, is left for developing software for new projects.

Thus, we have

$$PWF = n(t) + k(t) \qquad (16.2)$$

where PWF is the programming work force, n (t) is the normalized maintenance work force at time t, and k (t) is the normalized work force left for developing software for new projects at time t.

From Equation 16.2, we note that at $t = 0$, $n = 0$ and $PWF = k = 1$.

We define the fraction of the development force, x, assigned for maintenance at the completion of a software project as follows:

$$x = \frac{n}{PWF} \qquad (16.3)$$

However, at time $t = 0$, x is not defined.

If we start our first project at $t = 0$, then at its release, that is, $t = t_1$, we have

$$n(t_1) = x\,K(t_1) = x\,.\,1 = x \qquad (16.4)$$

and

$$K \equiv 1 - n = 1 - x \qquad (16.5)$$

After the release of the second project,

$$n = assignment\ to\ project\ no.1$$
$$+ assignment\ to\ project\ no.\ 2 \qquad (16.6)$$
$$= x + xK$$

By substituting Equation 16.5 into Equation 16.6, we get

$$n = x + x(1-x) \qquad (16.7)$$

and

$$K = PWF - n$$
$$= 1 - n \qquad (16.8)$$

Substituting Equation 16.7 into Equation 16.8 yields

$$d \equiv 1 - \left[x + x(1-x)\right]$$
$$= (1-x)^2 \qquad (16.9)$$

Similarly, using Equation 16.4 to Equation 16.9 for the mth release, we write

$$n = 1 - (1-x)^m \qquad (16.10)$$

and

$$K = (1-x)^m \qquad (16.11)$$

Example 16.1
After the completion of a software project, 20% of the work force is assigned to maintenance activities. There are a total of 8 projects of 2 years' duration. Estimate the percentage of the total work force that will be assigned to the maintenance activity of all the 8 projects.

By substituting the given values into Equation 16.10 and Equation 16.11, we get

$$n = 1 - (1 - 0.20)^8$$
$$= 0.8322$$

and

$$K = \left(1 - 0.20\right)^{0.8}$$
$$= 0.1678$$

Thus, about 83% of the entire work force will be assigned to the maintenance aspect of all eight software projects.

16.6 ROBOT MAINTENANCE REQUIREMENTS AND TYPES

The maintenance requirements of a robot are determined by the robot type and its application. Probably the most important part that affects the need for maintenance and the provision of maintenance is the robot power system.

Most robots used in the industrial sector can be grouped under two classifications: (a) electrical and (b) hydraulic with electrical controls [28]. Irrespective of robot type, the mechanical components of robots require careful attention.

Maintenance of robots used in the industrial sector can be divided into three basic categories as shown in Figure 16.1 [29]. These categories are preventive maintenance, corrective maintenance, and predictive maintenance. Preventive maintenance is concerned with the periodic servicing of robot system components. Corrective maintenance is concerned with repairing the robot to an operational state after its breakdown. Predictive maintenance is concerned with predicting failures that may occur and alerting the appropriate maintenance personnel. Many robots are equipped with sophisticated electronic components and sensors.

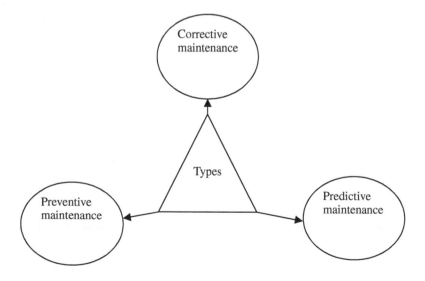

FIGURE 16.1 Basic types of maintenance for industrial robots.

16.7 ROBOT PARTS AND TOOLS FOR MAINTENANCE AND REPAIR

A robot is composed of various parts, subsystems, and accessories. Maintenance personnel must be familiar with such robot elements in order to perform their tasks effectively. Nonetheless, some of these elements are as follows [30]:

- Servo valve
- Hydraulic power supply
- Limit switch
- Cartesian coordinate system
- Cathode ray tube (CRT)
- Microprocessor
- Core memory
- Encoder
- Air cylinder
- DC servomotor
- Printed circuit board
- Bubble memory
- Mass memory device
- Strain gauge sensor
- Microcomputer
- Stepping motor
- Pressure transducer
- Alpha numeric keyboard
- Proximity sensor

In robot maintenance various types of tools are used, ranging from wrenches to diagnostic codes displayed on the robot control panel. Although the maintenance tools required are peculiar to the specific robot system in question, some of the most commonly used tools are as follows [31]:

- Torque wrenches
- Seal compressors
- Alignment fixtures
- Circuit card pullers
- Accumulator charging adaptors

16.8 ROBOT INSPECTION

Usually, robots are inspected regularly by their users. Nonetheless, the inspections of industrial robots may be grouped into two broad categories as shown in Figure 16.2 [32].

In category I some of the items checked prior to daily operations of the robots are [32]:

- The proper functioning of the emergency stop
- The proper working of the breaking device

FIGURE 16.2 Two broad categories of industrial robot inspections.

- The presence of abnormal noise
- The presence of abnormality in the robot supply air pressure
- The presence of abnormality in the robot supply oil pressure
- The presence of abnormal vibrations
- The presence of abnormality in the supply voltage
- The proper working of interlocking between the contact prevention equipment and the robot
- Damage to external electric wires and piping
- The presence of abnormality in the robot operation
- The proper working of interlocking the mechanism of associated items with the robot
- The state of items used for the prevention of contact with the robot in operation

In category II some of the items checked at regular intervals are [32]:

- The looseness of bolts in major robot parts
- Encoder abnormality
- Abnormal conditions in the electrical system
- Abnormality in the servo-system
- Abnormality in the power train
- Abnormal conditions in stoppers
- Abnormality in the operational troubleshooting function
- Abnormality in the lubrication of movable parts
- Abnormality in the air pressure system
- Abnormality in the oil pressure system

16.9 GUIDELINES FOR SAFEGUARDING ROBOT MAINTENANCE PERSONNEL

During robot maintenance utmost care must be given to protect robot maintenance and repair personnel. Four useful guidelines for this purpose are to [33]:

- Ensure that all maintenance personnel are properly protected from unexpected robot motion
- Ensure that all maintenance personnel have proper training in procedures appropriate to perform the required tasks safely

- Ensure that when a lockout or tag out procedure is not used, equally effective alternative safeguarding methods are employed.
- Ensure that the robot system is properly switched off during maintenance and repair activities as well as that the sources of power and the releasing of potentially dangerous stored energy are properly locked out or tagged

When it is not possible to turn off power during maintenance, some useful guidelines for protecting maintenance personnel are as follows [34]:

- Reduce the robot speed to a slow speed level.
- Place the robot arm in a predetermined position so the required maintenance tasks can be performed without exposing humans to trapping points.
- Make the emergency stop readily accessible and make restarting the robot impossible until the emergency stop device is reset through manual means.
- Place the entire control of the robot in the hands of the maintenance person.
- Place robot controls in the hands of a second person who is knowledgeable regarding potential robot-associated hazards and is capable of reacting fast to protect others in a moment of need.
- Use devices such as pins and blocks during maintenance to prevent the robot system's potentially hazardous movements.

16.10 MODEL FOR MAXIMIZING INCOME OF ROBOT SUBJECT TO REPAIR

This model is concerned with maximizing the income of a robot subject to failure and repair. The robot availability and unavailability are given by [29, 35]

$$AV_r(t) = \frac{\mu_r}{\lambda_r + \mu_r} + \frac{\lambda_r}{\lambda_r + \mu_r} e^{-(\lambda_r + \mu_r)t} \qquad (16.12)$$

and

$$UAV_r(t) = \frac{\lambda_r}{\lambda_r + \mu_r} \left[1 - e^{-(\lambda_r + \mu_r)t} \right] \qquad (16.13)$$

where λ_r is the robot constant failure rate, μ_r is the robot constant repair rate, AV_r (t) is the robot availability at time t, and UAV_r (t) is the robot unavailability at time t. For a large value of time t, Equation 16.12 and Equation 16.13 simplify to

$$AV_r = \frac{\mu_r}{\lambda_r + \mu_r} = \frac{MTTF_r}{MTTF_r + MTTR_r} \qquad (16.14)$$

and

$$UAV_r = \frac{\lambda_r}{\lambda_r + \mu_r} = \frac{MTTR_r}{MTTF_r + MTTR_r} \qquad (16.15)$$

where AV_r is the robot steady-state availability, UAV_r is the robot steady-state unavailability, $MTTF_r$ is the robot mean time to failure, and $MTTR_r$ is the robot mean time to repair.

Equation 16.14 and Equation 16.15 may also be interpreted as the fraction of the time the repair crew is idle and the fraction of the time the repair crew is working, respectively.

The robot maintenance crew cost per month is given by

$$C_r = \theta \mu_r = \frac{\theta}{MTTR_r} \qquad (16.16)$$

where θ is the robot maintenance cost constant that depends on the type of robot.

The expected income from the robot output per month is

$$I_e = I_f AV_r$$
$$= \frac{I_f (MTTF_r)}{MTTF_r + MTTR_r} \qquad (16.17)$$

where I_f is the income from the robot output per month if the robot worked full time.

Thus, the net income, NI_r, of the robot is

$$NI_r = I_e - C_r$$
$$= \frac{I_f (MTTF_r)}{MTTF_r + MTTR_r} - \frac{\theta}{MTTR_r} \qquad (16.18)$$

To maximize the net income of the robot, we differentiate Equation 16.18 with respect to $MTTR_r$ and set the resulting derivatives to zero:

$$\frac{d(NI_r)}{d(MTTR_r)} = -\frac{I_f (MTTF_r)}{(MTTF_r + MTTR_r)^2} + \frac{\theta}{(MTTR_r)^2} = 0 \qquad (16.19)$$

After rearranging Equation 16.19, we get

$$MMTR_r^* = \frac{MTTF_r}{\left[\frac{I_f(MTTF_r)}{\theta} \right]^{1/2} - 1} \qquad (16.20)$$

where $MTTR_r^*$ is the optimum value of $MTTR_r$.

By substituting Equation 16.20 into Equation 16.14, Equation 16.16, and Equation 16.18, respectively, we get

$$AV_r^* = 1 - \left[\frac{\theta}{I_f \, MTTF_r} \right]^{1/2} \qquad (16.21)$$

$$C_r^* = \left[\frac{\theta I_f}{MTTF_r} \right]^{1/2} - \frac{\theta}{MTTF_r} \qquad (16.22)$$

and

$$NI_r^* = I_f - \left[\frac{I_f \theta}{MTTF_r} \right]^{1/2} \qquad (16.23)$$

where $AV_r^*, C_r^*, and\ NI_r^*$ are the optimum values of AV_r, C_r, and NI_r, respectively.

16.11 MEDICAL EQUIPMENT CLASSIFICATION AND INDEXES FOR MAINTENANCE AND REPAIR

In the 1990s the Association for the Advancement of Medical Instrumentation (AAMI) performed a study with the objective of helping medical technology managers reduce repair and maintenance costs and improve the effectiveness of maintenance services [36]. In this study the AAMI classified the medical equipment into the following categories [7,36]:

Imaging and radiation therapy: Devices used for imaging patient anatomy and radiation therapy equipment. Some examples of such devices are linear accelerators, ultrasound devices, and x-ray machines.

Patient diagnostic: Devices connected to the patient and used to collect and analyze patient information. Devices such as endoscopes, physiologic monitors, and spirometers fall under this classification.

Life support and therapeutic: Devices that apply energy to the patient. Some examples of such devices are lasers, anesthesia machines, ventilators, and powered surgical instruments.

Patient environmental and transport: Patient beds and items used to transport patients or improve patient environment. Wheelchairs, gurneys, examination lights, and patient-room furniture fall under this classification.

Laboratory apparatus: Devices used in the preparation, analysis, and storage of *in vitro* patient specimens. Some examples of such devices are lab analyzers, centrifuges, and lab refrigeration equipment.

Miscellaneous medical equipment: This category includes items that are not included in the other five classifications. One example of such items is sterilizers.

The study also proposed and focused on the following three indexes for medical equipment repair and maintenance [7,36]:

16.11.1 INDEX 1

This is defined by

$$MT = \frac{TT}{n} \tag{16.24}$$

where MT is the mean turnaround time per repair, n is the total number of work orders or repairs, and TT is the total turnaround time.

This is a useful index that measures how much time elapses from a request from a customer until the failed item is put back in service.

16.11.2 INDEX 2

This is defined by

$$\alpha_r = \frac{C_s}{C_a} \tag{16.25}$$

where α_r is the cost ratio; C_a is the equipment, device, or item acquisition cost; and C_s is the service cost that includes the cost of items such as parts, labor, and materials for scheduled and unscheduled maintenance service.

The average values of α_r for the above six classifications of medical equipment are given in Reference 36. These are 5.1 for the laboratory apparatus classification, 3.5 for the life support and therapeutic classification, 5.6 for the imaging and radiation therapy classification, 4.4 for the patient environmental and transport classification, 2.6 for the patient diagnostic classification, and 2.6 for the miscellaneous medical equipment classification [7].

16.11.3 INDEX 3

This is defined by

$$RQ = \frac{TRR}{n} \tag{16.26}$$

where RQ is the number of repair requests completed per item, device, or equipment; n is the total number of devices, equipment, or items; and TRR is the total number of repair requests.

This index is quite useful to provide information concerning repair requests completed per device, equipment, or item. The average value of this index reported in Reference 36 is 0.8.

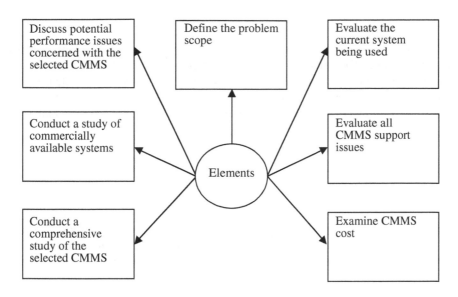

FIGURE 16.3 Major elements of the CMMS selection process.

16.12 COMPUTERIZED MAINTENANCE MANAGEMENT SYSTEM FOR MEDICAL EQUIPMENT AND ITS SELECTION

Computerized maintenance management systems (CMMSs) are often used by the clinical engineering departments of hospitals for collecting, storing, analyzing, and reporting data on repair and maintenance performed on medical devices and equipment. In turn, these data are used for purposes such as work order control, equipment management, quality improvement activities, reliability and maintainability studies, and cost control [7].

CMMSs for clinical engineering have become quite complex and sophisticated, and clinical engineering departments are finding them rather costly and time-consuming to develop, maintain internally, and update. However, today there are many commercially available CMMSs that can be used by hospital clinical engineering departments [7,37]. In selecting a commercial CMMS, the departments can follow a process similar to the prepurchase evaluation of medical equipment. The major elements of this process are shown in Figure 16.3 [37]. All these elements are discussed in detail in Reference 7.

16.13 MODELS FOR MEDICAL EQUIPMENT MAINTENANCE

Many mathematical models have been developed for application in general engineering equipment maintenance [16]. Some of these models can also be applied to medical equipment maintenance. This section presents two such models [7,38–39].

16.13.1 MODEL 1

This model determines the optimum number of inspections per equipment or device per unit of time. The determination of the optimum number of inspections is very important because inspection is often quite disruptive. However, it is quite useful to cut down on equipment downtime caused by unexpected breakdowns. This model minimizes the total downtime of the equipment to obtain the optimum number of inspections.

The total downtime of the equipment is given by

$$ETD = n\,T_{pi} + cT_{pb}n^{-1} \tag{16.27}$$

where ETD is the total downtime of the equipment per unit of time, n is the number of inspections per piece of equipment per unit of time, c is a constant for the specific facility under consideration, T_{pb} is the downtime per breakdown for the piece of equipment under consideration, and T_{pi} is the downtime per inspection for a piece of equipment under consideration.

By differentiating Equation 16.27 with respect to n and then setting the resulting expression equal to zero, we get

$$T_{pi} - cT_{pb} = 0 \tag{16.28}$$

By rearranging Equation 16.28, we get

$$n^* = \left[\frac{cT_{pb}}{T_{pi}} \right]^{1/2} \tag{16.29}$$

where n^* is the optimum number of inspections per piece of equipment per unit of time.

Example 16.2

Assume that for a certain piece of medical equipment we have the following data values:

- $c = 3$
- $T_{pi} = 0.06$ months
- $T_{pb} = 0.30$ months

Calculate the optimum number of inspections to be performed per month so that the medical equipment downtime is at minimum.

By substituting the given data values into Equation 16.29, we get

$$n^* = \left[\frac{(3)(0.30)}{0.06} \right]^{1/2}$$

$$= 3.87 \; \textit{inspections per month}$$

Thus, in order to keep the medical equipment downtime at a minimal level, the number of inspections to be performed is 3.87.

16.13.2 MODEL 2

This model is concerned with predicting the number of spares required. The number of spares needed for each item, equipment, and device in use is given by [38]

$$M_s = \lambda t + w \left[\lambda t \right]^{1/2} \tag{16.30}$$

where t is the mission time, M_s is the number of spares required, λ is the constant failure rate of the item under consideration, and w is associated with the cumulative normal distribution function. Its specific value is dependent upon a given confidence level for no stock out. Thus, for a specified value of confidence level, the value of w is obtained from the standardized cumulative normal distribution function table available in mathematical or other books [40–41], expressed by

$$P(w) = \frac{1}{\sqrt{2\Pi}} \int_{-\infty}^{w} e^{-x^2/2} \, dx \tag{16.31}$$

Example 16.2
Assume that the times to failure of a certain part used in an x-ray machine are exponentially distributed with the mean of 4,000 hours. Determine the number of the spare parts needed if the mission time is 3,000 hours and the confidence level for no stock out of the parts is 0.8643.

Thus, the part failure rate is

$$\lambda = \frac{1}{4,000} = 0.00025 \ failures/hour$$

For the given confidence level of 0.8643, by using the standardized cumulative normal distribution function table given in Reference 40, we get

$$w = 1.1$$

Using the above values and the data in Equation 16.30 yields

$$M_s = (0.00025)(3,000) + (1.1) \left[(0.00025)(3,000) \right]^{1/2}$$
$$\approx 2 \ parts$$

Thus, two spare parts are required.

16.14 PROBLEMS

1. Discuss the term *software maintenance*.
2. List at least six facts and figures concerning software maintenance.
3. What are the four types of software maintenance?
4. Discuss the following two methods with respect to software maintenance:
   - Impact analysis
   - Software configuration management
5. What are the basic types of maintenance for industrial robots?
6. List at least 15 important robot parts.
7. Discuss two broad categories of industrial robot inspections.
8. What are the main classifications of medical equipment?
9. Discuss three indexes considered useful for medical equipment maintenance and repair.
10. Assume that the times to failure of a certain part used in a piece of medical equipment are exponentially distributed with the mean of 2,500 hours. Calculate the number of spare parts needed if the mission time is 2,000 hours and the confidence level for no stock out of the parts is 0.8413.

REFERENCES

1. *IEEE Standard Glossary of Software Engineering Terminology*, IEEE-STD-610.12-1990, Institute of Electrical and Electronic Engineers, New York, 1991.
2. Omdahl, T.P., Ed., *Reliability, Availability (RAM) Dictionary*, ASQC Quality Press, Milwaukee, WI, 1988.
3. Boem, B.W., *Software Engineering Economics*, Prentice Hall, Englewood Cliffs, NJ, 1981.
4. Stevenson, C., *Software Engineering Productivity*, Chapman and Hall, London, 1995.
5. Martin, J., *Fourth-Generation Languages*, Vol. 1, Prentice Hall, Englewood Cliffs, NJ, 1985.
6. *Maintainability Engineering Theory and Practice*, AMC Pamphlet No. AMCP 706-133, prepared by the Department of the Army, Headquarters U.S. Material Command, Alexandria, VA, 1976.
7. Dhillon, B.S., *Medical Device Reliability and Associated Areas*, CRC Press, Boca Raton, FL, 2000.
8. Stacey, D., *Software Engineering*, Course 27-320 Lecture Notes, Department of Computers Science, University of Guelph, Guelph, Ontario, Canada, 1999.
9. Charrette, R.N., *Software Engineering Environments*, Intertext Publications, New York, 1986.
10. Coleman, D., Using metrics to evaluate software system maintainability, *Computer*, 27(8), 44–49, 1997.
11. Boeing Company, *Software Cost Measuring and Reporting*, ASD, Document No. D180-22813-1, U.S. Air Force, Washington, DC, 1979.
12. Pfleeger, S.L., *Software Engineering Theory and Practice*, Prentice Hall, Upper Saddle River, NJ, 1998.
13. Leintz, B.P. and Swanson, E.B., Problems in application software maintenance, *Communications of the ACM*, 24(11), 763–769, 1981.

14. Holbrook, H.B. and Thebaut, S.M., *A Survey of Software Maintenance Tools that Enhance Program Understanding*, Report No. SERC-TR-9-F, Software Engineering Research Center, Department of Science, Purdue University, West Lafayette, IN, 1987.
15. Pfleeger, S.L. and Bohner, S., A framework for maintenance metrics, *Proceedings of the IEEE Conference on Software Maintenance*, 1990, pp. 225–230.
16. Dhillon, B.S., *Engineering Maintenance: A Modern Approach*, CRC Press, Boca Raton, FL, 2002.
17. Schneider, G.R.E., Structural software maintenance, *Proceedings of the AFPIS National Computer Conference*, 1983, pp. 137–144.
18. Arthur, L.J., *Software Evolution: The Software Maintenance Challenge*, John Wiley & Sons, New York, 1983.
19. Gilb, T., *Principles of Software Engineering Management*, Addison-Wesley, Wokingham, Berkshire, U.K., 1988.
20. Hall, R.P., Seven ways to cut software maintenance costs, *Datamation*, July, 81–87, 1987.
21. Lindhorst, W.M., Scheduled maintenance of application software, *Datamation*, May, 64–67, 1973.
22. Parikh, G., Three Ts keys to maintenance programming, *Computing S.A.*, April 22, 19, 1981.
23. Yourdon, E., Structured maintenance, in *Techniques of Program and System Maintenance*, Parikh, G., Ed., Ethnotech, Lincoln, NE, 1980, pp. 211–213.
24. Dhillon, B.S., *Life Cycle Costing*, Gordon and Breach Science Publishers, New York, 1989.
25. Sheldon, M.R., *Life Cycle Costing: A Better Method of Government Procurement*, Westview Press, Boulder, CO, 1979.
26. Shooman, M.L., *Software Engineering*, McGraw-Hill, New York, 1983.
27. Mills, M.D., Software development, *Proceedings of the IEEE Second International Conference on Software Engineering*, Vol. II, 1976, pp. 79–83.
28. Lester, W.A., Lannon, R.P., and Bellandi R., Robot users need to have a program for maintenance, *Industrial Engineering*, January, 28–32, 1985.
29. Dhillon, B.S., *Robot Reliability and Safety*, Springer Verlag, New York, 1991.
30. Ottinger, L.V., Robot system's success based on maintenance, in *Robotics*, Fishers, E.L., Ed., Industrial Engineering and Management Press, Institute of Industrial Engineers, Atlanta, GA, 1983, pp. 204–208.
31. Munson, G.E., Industrial robots: reliability, maintenance, and safety, in *Handbook of Industrial Robotics*, Nof, S.Y., ed., John Wiley & Sons, New York, 1985, pp. 722–758.
32. *An Interpretation of the Technical Guidance on Safety Standards in the Use, etc., of Industrial Robots*, Japanese Ministry of Labor, Ed., Japanese Industrial Safety and Health Association, Tokyo, 1985.
33. *American National Standard for Industrial Robots and Robot Systems: Safety Requirements*, Document No. ANSI/RIA R15-6, prepared by the Robotic Industries Association, Ann Arbor, MI, 1986.
34. Lodge, J.E., How to protect robot maintenance workers, *National Safety News*, June, 48–51, 1984.
35. Dhillon, B.S., *Design Reliability: Fundamentals and Applications*, CRC Press, Boca Raton, FL, 1999.
36. Cohen, T., Validating medical equipment repair and maintenance metrics: a progress report, *Biomedical Instrumentation and Technology*, January/February, 23–32, 1997.

37. Cohen, T., Computerized maintenance management systems: how to match your department needs with commercially available products, *Journal of Clinical Engineering*, 20(6), 457–461, 1995.
38. Ebel, G. and Lang, A., Reliability approach to the spare parts problem, *Proceedings of the Ninth National Symposium on Reliability and Quality Control*, 1963, pp. 85–92.
39. Wild, R., *Essentials of Production and Operations Management*, Holt, Rinehart and Winston, London, 1985.
40. Spiegel, M.R., *Mathematical Handbook of Formulas and Tables*, McGraw-Hill, New York, 1968.
41. Kales, P., *Reliability*, Prentice Hall, Inc., Upper Saddle River, NJ, 1998.

Index

Milton Keynes UK
Ingram Content Group UK Ltd.
UKHW040102071024
449327UK00019B/754